Praise for *Inside Animal Hearts and Minds*

"Belinda Recio has written a fascinating account of animal emotions and animal intelligence. She makes the stories she tells available to a wide audience, and her examples are fair, friendly, and charming. I cannot imagine any animal lover not finding this a wonderful book!"

—Jeffrey Moussaieff Masson, bestselling author of nine books on the emotional life of animals, including *Beasts: What Animals Can Teach Us About the Origins of Good and Evil*

"In an increasingly human-dominated world, nonhuman animals need all the help they can get. Belinda Recio's *Inside Animal Hearts and Minds* will make it easier for people to re-wild their hearts, and reconnect with other animals. It's essential that as we learn about the inner lives of animals—their cognitive, emotional, and moral capacities—we take action on their behalf. Ms. Recio's book could be a game changer."

—Marc Bekoff, professor emeritus of Ecology and Evolutionary Biology at the University of Colorado, Boulder, and author of numerous books on animals, including *The Emotional Lives of Animals: A Leading Scientist Explores Animal Joy, Sorrow, and Empathy and Why They Matter*

When Animals Rescue

Amazing True Stories about Heroic and Helpful Creatures

Belinda Recio

Foreword by Mark Rowlands

Skyhorse Publishing

Skyhorse Publishing books may be purchased in bulk at special discounts
for sales promotion, corporate gifts, fund-raising, or educational
purposes. Special editions can also be created to specifications. For
details, contact the Special Sales Department, Skyhorse Publishing,
307 West 36th Street, 11th Floor, New York, NY 10018 or
info@skyhorsepublishing.com.

Skyhorse® and Skyhorse Publishing® are registered trademarks of
Skyhorse Publishing, Inc.®, a Delaware corporation.

Visit our website at www.skyhorsepublishing.com.

10 9 8 7 6 5 4

Library of Congress Cataloging-in-Publication Data is available on file.
Library of Congress Control Number: 2020933029

Cover design by Erin Seaward-Hiatt
Cover photo credit: iStockphoto

Print ISBN: 978-1-5107-6260-2

Printed in China

Contents

For Spooner,
whose charisma, intelligence, sense of humor, and love
enriched my life beyond measure.

The heart has reasons of which
reason knows nothing.

Blaise Pascal, *Pensées*

Foreword

Arthur Schopenhauer, popularly known as the great philosopher of pessimism, once encapsulated the human moral predicament in a way that, to my ears at least, almost sounds optimistic.

> From this point of view, we might well consider the proper form of address to be not *Monsieur, Sir, Mein Herr*, but *my fellow sufferer, Soci Malorum, compagnons de misère*. This . . . reminds us of that which is after all the most necessary thing in life—the tolerance, patience, regard, and love of neighbor of which everyone stands in need and which, therefore, every man owes his fellow.

This world is everything we have, and we are therefore, I suppose, obliged to love it. Nevertheless, we know that, in certain respects, it is a bad world. It is not merely that it is a world replete with suffering: pain, torment on an almost unimaginable scale. Worse: this suffering is a consequence of the basic design principles of the universe. The first law of thermodynamics tells us that energy can neither be created nor destroyed. The second law tells us that the disorder of any complex structure will increase in the absence of an input of energy. Taken together, these laws entail that destruction and death are built into our universe, as simple consequences of its design. To remain in existence, complex structures—such as you and I—must break down other complex structures and consume their energy. Even worse: when the light of consciousness finally evolved in this formerly dark universe, destruction was translated into suffering. And, worst of all: it is overwhelmingly likely that in lives such as ours, and in the lives of other conscious creatures, suffering will predominate. Suffering is a sign that the struggle to stay in existence is going badly, and you always have to pay special attention to this. Happiness—the sign of a struggle going well—you can afford to ignore. Suffering you have to deal with. And so, Schopenhauer concluded, we are, and must always be, more attuned to suffering than happiness. Unless we are very, very lucky, the suffering in our lives will outweigh the happiness. This is our predicament. His answer, his moral categorical imperative, was compassion.

In such a world, Schopenhauer's vision may seem unlikely. How could the sorts of things to which Schopenhauer appealed—tolerance, patience, regard, love, compassion—exist in a world designed in this way? How could goodness—moral goodness—evolve in a world predicated on destruction? Put it this way: if you were told beforehand that *this* is how the universe was going be designed—by way of two laws that had such baleful consequences—how much money would you have been willing to put on the emergence of these sorts of qualities? I have to admit: I suspect I wouldn't have seen it coming.

Yet it did. These qualities emerged. As Belinda Recio's wonderful book demonstrates, they are not at all the exclusive preserve of humans—as we self-servingly try to convince ourselves—but are distributed widely throughout the animal kingdom. We humans are not the *only* moral animals. Indeed, we are probably not even the *most* moral of animals. That a universe designed along the most seemingly malicious lines produced these qualities, the kinds of behaviors documented in this book, seems almost *miraculous* to me. Sometimes I wish I were a more religiously inclined man, for then I could use the word *miracle* without compunction or scruple. Without it seeming a little *odd*. But, anyway, it is not a miracle. None of it is. There is no supernatural agency. There are no violations of the laws of nature. And we have a reasonable understanding of why and how goodness of this sort arose, in other animals as well as ourselves. Nevertheless, there is still something about it all that just seems so fantastically improbable: that a universe designed this way should have produced animals who do the things described here. Our home being what it is, this is what we might think of as a miracle of improbability.

In this book you meet animals who prove themselves to be so much more than the biological marionettes, or stimulus-response machines, that we have, for so long, tried to convince ourselves that they are. You will meet animals—little miracles of improbability—who are tolerant, patient, loving, kind, and compassionate. Not to mention intelligent and brave. You will meet animals who are our friends, our fellow travelers, our fellow sufferers, our *compagnons de misère*. You will, above all, meet animals who are *good*. If I understand Schopenhauer at all, then I know he would have approved.

—Mark Rowlands
Miami, Florida
March 2020

Acknowledgments

I am grateful to all the scientists and animal caregivers whose work and observations made this book possible.

Much gratitude and appreciation to Mark Rowlands for writing the foreword.

Thank you to my editor, Kim Lim, for encouraging me to write this book. A warm thank-you to my family and friends who once again listened to all the animal stories, shared in my amazement, and supported me in various ways. Special thanks to Joan Parisi Wilcox for her help throughout this project. And my deepest gratitude to my husband, Ed Blomquist, for all the ways he supports my work; and to our dog, Spooner, who regularly rescued me from my desk—and sometimes from myself—for twelve wonderful years.

Author's Notes

Every effort has been made to credit the scientists, researchers, and writers whose work I present herein. I regret any omission and pledge to correct errors called to my attention in subsequent editions.

Regarding terminology: although humans are animals, I use the term *animal* to refer to nonhuman animals (except in the Introduction). In order to avoid objectifying animals, instead of *it* I use the pronouns *him* and *her*. Finally, I use the term *owner* despite how this designation degrades the role of animals in our lives to that of property. Animals who are under our care are more like family members, and our roles are closer to that of guardians rather than owners. But for the sake of readability, I have reluctantly defaulted to the common term, *owner*.

Introduction

When my editor first approached me and asked me to consider writing a book on animals who rescue, protect, or otherwise help out a human or other animal, I was hesitant. I had heard a few stories about animals engaging in these kinds of activities, but I wasn't sure there would be enough verified content to fill a book. But as it turned out, there are plenty of true stories about animals behaving in remarkably kind and helpful ways.

I wasn't very far into my research before I remembered a quote that helped me decide to write this book. In her essay "A Writer in the World," author and naturalist Annie Dillard wrote, "You were made and set here to give voice to this, your own astonishment." This quote came to mind because the stories I was finding were truly astonishing—humpback whales protecting a biologist from a shark, wild gorillas working together to dismantle poachers' snares, seals who saved a woman from drowning, and many others. If the animals in these stories had been human, they would have been considered charitable, compassionate, or even heroic. Their acts of kindness—even if unintentional—remind us of what is best about the world. So I decided to take Dillard's advice and write this book to give voice to my astonishment.

But what surprised me the most was the fact that I was astonished at all. This is the second book I have written about animal behavior, and I have covered this subject as a journalist for more than a decade, so I wasn't expecting to feel surprise. In my first book on animal behavior, *Inside Animal Hearts and Minds: Bears That Count, Goats That Surf, and Other True Stories of Animal Intelligence and Emotion*, I wrote about ants who recognize themselves in a mirror, magpies who appear to hold funerals, parrots who name their chicks, apes who engage in make-believe, and much more. So why did it amaze me to learn that a parrot valiantly fought off an attacker to protect his owner or that a wild monkey tried to revive an unresponsive companion using CPR—chest compressions and mouth-to-mouth resuscitation? Why, after learning so much about animals, was I still struck by how human-like they seem? And why was I still thinking of these behaviors as *human-like* rather than *parrot-like* or *monkey-like*?

Writing this book made me realize that the belief in human exceptionalism is deeply entrenched—even in someone like myself, who writes about animals. The idea that humans are so very different from other animals extends deep into the past and is primarily rooted in various religious and philosophical traditions. But it continues into the present, in a myriad of ways, including scientists' long-standing disapproval of anthropomorphism—the tendency to attribute human traits to other species. Within the general scientific community, it has been considered poor methodology to try to understand animal behavior by relating it to human behavior, because doing so may lead to misinterpretation of evidence and data. And there's no doubt that it's important to try to understand animals on their own terms, and not solely on the basis of how similar to humans they might seem. We need to be careful not to let our excitement about our commonalities prevent us from seeing them as they truly are.

However, completely dismissing behavioral similarities between human and animal behavior only makes sense if we believe that there isn't any *emotional* continuity between humans and animals. Intuitively, this just doesn't ring true. Many people who live with animals believe that they experience a range of emotions similar to those experienced by humans, and dozens of recent studies validate this belief. Because animal emotions are manifested in behaviors that can be observed and measured, scientists have been able to determine that a wide variety of animal species—ranging from mice to whales—appear to feel many of the same emotions that humans do. In fact, after decades of research, primatologist Frans de Waal goes one step further and asserts that "there are no human emotions that don't have a parallel in animal emotions."

To help make his point about how animals share the emotional spectrum with humans, de Waal compares emotions to biological organs. "Like organs, the emotions evolved over millions of years to serve essential functions. Their usefulness has been tested again and again, giving them the wisdom of ages. They nudge us to do what is best for us. Some emotions may be more developed in humans, or apply to a wider range of circumstances, but none is fundamentally new." So, just as both humans and animals have hearts, brains, stomachs, and other organs (although these organs differ somewhat from species to species), they also appear to experience love, joy, grief, and other emotions (although the extent and expression of these emotions may differ as well).

With this emotional continuity between humans and animals in mind, we should stop thinking that a hippopotamus who saved a wildebeest from a crocodile was behaving in a "human-like" way. Whatever emotion might motivate a hippopotamus to behave this way—maternal protectiveness, empathetic concern, or something else—the emotion is one that is likely experienced by many animals, not just by humans.

So as you read the following stories—about animals acting in ways that seem motivated by empathy, compassion, altruism, and other emotions that we associate with our highest human virtues—I hope you will consider the possibility that what's best about our human natures just might be our *animal* natures.

Chapter 1
Whale

A Beautiful Question

When the roughly forty-seven-foot-long, fifty-thousand-pound humpback approached marine biologist Nan Hauser—who was swimming near the whale—she assumed that he would swim around her. Instead, the massive whale started to push her around. He nudged her with his rostrum (snout) and tried to lift her out of the water onto his head, belly, and back. He pushed and pulled her with his huge pectoral fin and even tried to tuck her under it. In all of her twenty-eight years of swimming with whales, Hauser had never experienced a whale handling her the way this humpback was.

As the whale continued to graze and grope her, Hauser grew increasingly worried about her safety. She knew that if the whale bumped her too hard, he could very easily break her bones or cause internal injuries; and if he held her under his fin too long, she could drown. Even though the biologist believed that her life was in danger, she tried to stay calm and kept her underwater camera rolling. After about ten of the longest minutes of her life, she was finally able to disengage from the whale and swim toward the safety of her research vessel. Other than some scrapes from the barnacles on the whale's body, Hauser was physically okay. But mentally, she was utterly perplexed by the whale's behavior.

When Hauser had just about reached the boat, she looked back and saw a second humpback tail-whacking what appeared to be yet another—a third—whale. Just as this third whale started swimming toward her, she suddenly saw something that confirmed her life really had been in danger: what she had thought was a third whale was actually a fifteen-foot-long tiger shark. The humpback hadn't been attacking her—he had been protecting her from the shark. And while her humpback hero was trying to keep her away from the shark, the second humpback had been doing his part by tail slapping it.

Once she was back on her research vessel, Hauser and others on board reviewed the video footage of the encounter and confirmed that the humpbacks appeared to be trying to protect her from the shark.

Hauser is president of the Cook Islands–based Center for Cetacean Research and Conservation in the South Pacific, and she has nearly three decades of experience researching whales. She had heard other stories of humpbacks protecting marine mammals from orcas and sharks, but as far as she knew, this was the first time a humpback had protected a human. When talking about her encounter, Hauser likes to point out that she had spent twenty-eight years protecting whales before she realized that they might have been protecting her, too.

Four days after the humpbacks protected her from the shark, Hauser was in her boat when she saw a humpback surfacing to check her out. Hauser entered the water with her camera, and the whale swam up underneath her until they were only four feet apart. The whale then placed both pectoral fins around her, in what felt to Hauser "like a hug."

When asked in an interview with National Public Radio why a whale would behave altruistically, Hauser answered, "I've been studying humpbacks for twenty-eight years, and I plan on spending a lot more of my life trying to figure that question out because it's actually a beautiful question to try to answer."

A Compassionate Instinct?

The seal was floating on a raft of ice when marine ecologist Robert Pitman noticed her. A second later, Pitman saw a group of orcas working together to create a wave that knocked the seal into the water. Things looked pretty bad for the seal—until a pair of humpbacks arrived. The seal swam directly toward the humpbacks, and as she approached the whales, a wave lifted her right up onto the chest of one of the whales, who was floating on his back. Pitman assumed that being out of the water and safe on the humpback's chest would only bring the seal a few

Orcas will sometimes beach themselves on land or jump onto an ice floe in order to catch a seal.

seconds of reprieve, because once she slipped back into the sea, the orcas would finally get their dinner.

Then the strangest thing happened. As the seal started to slip off the whale's chest, the whale gently nudged her back on with his flipper, keeping her away from the orcas. Later, after the marauding orcas had left, the seal slid off the whale's chest and made her way to the safety of another island of ice.

This episode happened while Pitman and a team of scientists were in Antarctica researching orcas. The BBC had hitched a ride on their research vessel so that they could capture footage for a documentary, *Frozen Planet*. A few days before he saw the humpbacks rescue the seal, Pitman and the others on board saw orcas interacting with humpbacks. At that time, they thought that the humpbacks might be under attack. They moved closer to get a better look, but nothing much seemed to be going on between the humpbacks and orcas, and soon both species dispersed. But later, when they looked at the footage shot by the BBC film crew, they saw that a Weddell seal had been taking refuge between two humpbacks.

If that was all Pitman had seen, he might have written it off as just the seal's good luck at having found a couple of humpbacks to hide between. But when he witnessed the second rescue, he started to wonder if something interesting was going on with humpback whales. So he reached out to others who regularly observe humpbacks and asked if they had ever seen them come to another animal's defense. Pitman received 115 descriptions of similar encounters, many documented by photographs and videos. In 90 percent of those encounters in which the animal under attack could be identified, it was not another humpback, so it was clear the humpbacks were not simply defending one of their own.

No one knows why these gentle giants appear to come to the rescue of other animals, but it might have something to do with their gray matter. Humpback brains contain spindle neurons—cells that, in the human brain, are associated with the ability to feel empathy. So it's possible that humpbacks are "wired" for feeling and understanding the emotions of others.

Beluga to the Rescue

Almost everyone has an idea of what their dream job would be, and for Yang Yun, it was being a whale trainer at China's Polar Land Aquarium. In 2009, Yun had a shot at this career, but applying for it involved more than just submitting a résumé. Applicants had to participate in a free-diving competition, demonstrating how far they could dive into the twenty-foot-deep arctic pool that housed the beluga whales and how long they could stay underwater. The applicant who dove the deepest and lasted the longest would get the job.

Yun signed up to try out, and everything went well at first. But then, as she descended to about fifteen feet under the icy water, her legs stopped working. The arctic temperature of the water caused her muscles to cramp, preventing her from swimming to the surface. She panicked and began to choke, which only caused her to drop farther into the pool. Just as the aspiring whale trainer gave up hope and believed she was going to die, she suddenly felt herself being pushed to the surface.

Belugas are social, gregarious animals who migrate, hunt, and play together. In the wild, they exhibit curiosity toward humans and often approach boats and divers.

It turns out that two of the belugas in the pool—Nicola and Mila—had noticed Yun's distress. Mila immediately took matters into her own . . . mouth. She gently grasped Yun's leg in her mouth and pushed her up to the surface of the pool. Yun survived the ordeal without injury, profoundly grateful for Mila's immediate action, which had saved her life. The spokesperson for the aquarium said that Yun was especially lucky because Mila had recognized that Yun needed help before any of the aquarium staff did, which saved precious time.

Why did Mila rescue Yun? It might be because belugas are social animals who hunt together and constantly communicate with one another. As a result of their social behavior, they develop a keen sensitivity to their pod mates. It's possible that Mila, who was socialized with humans, had developed a sensitivity to people as well and so responded to Yun's distress as if she were a pod mate.

A Whale of Gratitude

Given that whales behave altruistically toward other species, is it possible that they might be capable of recognizing—and even appreciating—altruistic behavior directed toward themselves? The divers who came to the rescue of a humpback believe they know the answer to this question.

James Moskito was one of a team of divers called in to try to rescue a humpback whale entangled in commercial crab traps near the Farallon Islands, about thirty miles west of San Francisco, in December 2005. He was the first to approach the whale, and when he saw the extent of the whale's entanglement, his heart sank. She was floating near the surface of the sea, surrounded by buoys, each of which was tied with numerous weighted lines to about a dozen ninety-pound metal crab traps sitting on the ocean floor, roughly 250 feet below. The lines were so deeply embedded into the whale's flesh that many were no longer visible. Moskito didn't believe that the whale had a chance, but he was determined to do what he could.

The Cells That Make Us . . . Whale?

Another explanation for Mila's compassionate behavior could be found in the cells in her brain. Beluga brains, like those of humpbacks and other whales, contain spindle neurons, which are associated with self-awareness, a theory of mind (awareness of other minds), social intelligence, and empathy. The spindle neuron has often been described as the "cell that makes us human." But this belief changed in 2006 when two scientists, Patrick Hof and Estel Van Der Gucht, discovered spindle cells in the brains of humpbacks and other species of whales. Not only do whales have spindle cells, but they may have three times as many as humans do, even after adjusting for brain size. Even more interesting, whales may have had this kind of cell for twice as long as humans, evolutionarily speaking. So perhaps instead of thinking that spindle cells—and all the cognitive functions they enable—are what make us human, we could think of them as the cells that "make us whale."

Within minutes, Moskito was joined by another diver, Tim Young, and the two snorkeled around the distressed humpback to further evaluate the situation. They saw crab lines tightly wound around the whale's pectoral fin, head, and mouth. When Moskito dove under the surface, he saw that her tail was tangled in the weighted crab-trap lines, which were pulling her downward, forcing her to struggle to keep her blowhole out of the water.

Although scientists do not know if animals feel gratitude, there are many stories about animals showing pleasure after a human treats them kindly. When the animal on the receiving end of a kindness directs their pleasure toward the individual who treated them kindly, it seems reasonable to consider the possibility that the animal is expressing something akin to gratitude.

The two divers switched from snorkels to scuba gear and returned to the water with curved saws. They decided to try to cut through the lines wrapped around her pectoral fin, even though they realized they could be seriously injured if the whale slapped her massive fin while they were cutting. But as soon as they started, the whale stopped moving and remained calm. She watched the divers and seemed to cooperate, appearing to understand that they were there to help.

After freeing her fin, Young next removed the lines that were embedded in the whale's mouth and head. At the other end of the whale, under many feet of water, Moskito worked tirelessly to cut the lines that entangled her tail. Two more divers joined Young and Moskito, removing other lines.

Finally, after five hours of working on the lines, the divers finally freed the whale. She immediately dove under the surface and disappeared. Or so Moskito thought. A moment later he saw the whale swimming straight at him, and his adrenaline spiked as he anticipated the worst. But the humpback did not charge him. Instead she came to an abrupt stop a few inches away from his chest and gently nudged him with her rostrum (snout), looked at him with one of her massive eyes, and swam next to him, gently brushing up against him. Moskito stroked her as she swam in circles. She also approached the other divers, nudging and nuzzling them, too. Moskito described the whale's behavior as "affectionate, like a dog that's happy to see you."

As the whale swam around them, Moskito saw a line in her month that they had missed. He waited for an opportunity to cut it, and as he did, the whale made a deep humming sound. When that last line was removed, the whale circled around all of the divers a few more times, gradually widening her circles and moving away.

Although Moskito and the other divers cannot prove the whale was expressing gratitude for being released from what would have been a fatal tangle of lines, they nonetheless believe that the whale's behavior was her way of saying, "Thank you."

Chapter 2
Parrot

More than Just a Mimic

Alex was an African grey parrot who was the focus of one the most ambitious animal cognition studies in history. Scientist Irene Pepperberg worked with Alex for thirty years, during which she taught him more than one hundred words. But Alex didn't merely "parrot" these words—he used them correctly and creatively.

One especially amusing example of Alex's verbal prowess involves his love of being tickled around his neck and a toy parrot he received as a gift. Alex would often approach Pepperberg and her research assistants, bend his head to expose the back of his neck, and say, "You tickle." Pretty much everyone obliged and tickled him. As for the toy parrot, Alex didn't seem to like it at first and kept his distance. But about a week after he received the toy, Alex appeared to warm up to it. He approached it, bent his head, and uttered his "You tickle" command. When the toy parrot failed to tickle him, Alex looked at it and said, "You turkey."

Alex had learned the insult from research assistants who often called him a turkey when he didn't respond the way he was supposed to. No one had taught Alex to use the word "turkey" as a reprimand for noncompliance; he had figured it out by himself. This incident, along with many other examples of Alex's appropriate and contextual use of words, makes a strong case that Alex understood the meaning and impact of the words he uttered.

As special as Alex was, many parrot owners claim that their parrots also use language intentionally. For example, Willie, a Quaker parrot (also known as a monk parakeet), not only appeared to use words purposefully, but once potentially saved a life by using human words with apparent intention. Prior to the incident, Willie had learned to mimic a few fun phrases, such as "Come here," "I want out," "I love you," "Give me a kiss," and a few expletives that he wasn't meant to mimic. He also learned to say "Mama," which was his name for Meagan Howard, his owner.

In November 2008, Meagan was living in Denver with Willie, her roommate Samantha, and Samantha's two-year-old daughter, Hannah. One day Meagan offered to babysit Hannah while Samantha went out for the day. After Samantha left, Meagan toasted a pastry for Hannah's breakfast but then determined that it was too hot for Hannah to eat. So, Meagan set the pastry on the dining table and distracted Hannah with a television program while it cooled.

Monk parakeets are intelligent birds who often develop large vocabularies of words when living in captivity. Many people who live with monk parakeets claim that their birds sometimes spontaneously string words together to form meaningful phrases that they use in appropriate contexts.

Once the toddler was engaged with the program, Meagan decided to take a quick restroom break. Almost immediately after leaving Hannah, Meagan heard Willie making a commotion in his cage. He was flapping his wings and frantically vocalizing in a way that Meagan hadn't heard before. But he wasn't simply squawking. He was shouting two words over and over, "Mama, baby! Mama, baby!"

Meagan rushed back into the room to find Hannah choking on the pastry and gasping for air, her face and lips already blue. Meagan acted quickly and performed the Heimlich maneuver, which cleared the toddler's throat and allowed her to breathe normally again. It was only after Hannah recovered that Willie calmed down.

Meagan was used to Willie calling out "Mama," but she was surprised that he had said "baby," a word he hadn't spoken until that morning. Willie had undoubtedly heard his human roommates say "baby" frequently, and he had clearly learned to mimic it, which parrots often do. But what stunned Meagan was that Willie had combined "mama" with "baby," and then went out of his way—by flapping his wings and squawking his message with such urgency—to get her attention when it was so desperately needed.

We don't know what motivated Willie to call out at just the right moment. He might have been simply responding to Hannah's distress, which scientists call "emotional contagion." Animals will sometimes demonstrate this basic kind of empathy by responding to another's mood. Sometimes the reacting animal "catches" and displays the fear, or other emotion, exhibited by the emoting animal. Other times, the reacting animal might try to console the distressed animal. But the timing of Willie's reaction, combined with his use of appropriate words, suggests that Willie wasn't merely experiencing emotional contagion. He seemed to understand that Hannah was in distress and that Meagan's attention was needed. Further evidence supporting the idea that Willy used words purposefully is Meagan's report that he never again uttered that combination of words.

Samantha was deeply grateful to both Meagan and Willie for saving her daughter's life. She felt especially indebted to Willie because if he hadn't alerted Meagan, the story could have ended tragically. Meagan felt the same way about Willie and described him as "the real hero." And they weren't the only ones who felt that way. At a Breakfast of Champions event attended by the governor of Colorado, the mayor of Denver, and many others, Willie was presented the Red Cross Animal Lifesaver Award for his heroic act.

Need a Guard Dog? Consider a Parrot Instead!

Alex, the African grey parrot who worked with Dr. Irene Pepperberg, not only helped the world to reconsider avian intelligence, he also brought a lot of attention to his species. African greys are known to be uncannily intelligent, but what is even more incredible about them is their emotional capacity, which often results in these parrots forming deep attachments to their owners. In fact, people who live

Recent research shows that African grey parrots will readily help one another, even if there's nothing in it for themselves.

with African greys will often describe themselves as being owned by the parrot rather than the other way around.

Rachel Mancino shared a deep bond with her female African grey, Wunsy. Knowing that captive birds need exercise, Rachel took Wunsy for a daily walk. Wunsy would ride on Rachel's shoulder—tethered to her wrist on a lead—as they strolled through the park. Wunsy was able to fly just above Rachel when she felt like it, or she could simply perch on Rachel's shoulder and go along for the ride.

When Wunsy was only about eight months old (parrots can live to be eighty years old in captivity, so eight months is still very young), she was out for a walk with Rachel in a north London park. Shortly into their walk, Rachel saw—through her peripheral

The Human-Parrot Bond

Parrot enthusiasts often claim that the bond they share with these intelligent, playful, intuitive, and affectionate birds is as deep as any they have shared with dogs, cats, and other companion animals. This bond might arise from their ability to imitate human speech and use words meaningfully. But many ornithologists believe that it isn't just parrots' mimicry skills that gives rise to deep bonds with people; instead it's their innate tendency to bond with one another. In the wild, parrots are devoted mates and caring flock members and will not leave one of their own when ill or injured. When another parrot is not available, parrots transfer this behavior to their human caregivers instead, becoming protective and devoted companions.

A poignant example of parrot devotion can be found in the book *The Parrot Who Owns Me: The Story of a Relationship* by ornithologist Joanna Burger. Burger shares the story about how her red-lored Amazon parrot, Tiko, would "protect" her by standing guard over her when she was working, vocalizing to warn her of any hawks he saw outside her window. But Tiko didn't just try to protect Burger, he also wouldn't leave her side when she fell ill. Once, when she was bedridden for six weeks, Tiko was so insistent on staying close to her—nuzzling her and preening her hair—that Burger's husband had to bring Tiko's food to the bedroom to get him to eat.

vision—a man walking quickly toward her. As he followed close behind, Rachel worried for a second, but then dismissed her concern. Wunsy, however, appeared to have a different assessment of the situation. She turned herself around on Rachel's shoulder so that she faced the stranger. Less than a minute later, the man jumped on Rachel, grabbed her neck, and tried to push her to the ground. Like a devoted guard dog, Wunsy wasted no time. She squawked loudly and swooped on the assailant, slapping him in the face with her wings. Luckily, Wunsy's defensive behavior caused the man to abandon his attack and run off, but not before looking back, presumably to see if Wunsy was in pursuit.

Rachel later discovered that the man who had assaulted her had attacked someone else just minutes earlier. She realized that Wunsy had probably saved her from a mugging, or possibly even worse.

When speaking to the BBC about the incident, Rachel jokingly said that because Wunsy is "still a baby, she's still mastering her coming-to-the-rescue technique." However, despite her age, Wunsy had behaved like an astute guard bird. She seemed to sense that the man was a threat and had repositioned herself in order to watch him, and perhaps even to be in a better position to defend Rachel if necessary.

In retrospect, Wunsy might have better assessed the potential threat than Rachel had.

Wunsy isn't the only parrot who attracted the attention of the press for behaving more like a guard dog than a bird. A Fort Smith, Arkansas, macaw named Charlie protected his owner from two intruders who broke into his home. One day in September 2011, Jack Dukes answered his door and let two men, who presented themselves as neighbors in need of help, into his apartment. But they weren't neighbors— they quickly made it clear that they were there to rob Jack and began knocking him around. Like Wunsy, Charlie wasted no time and started squawking and attacking the intruders, even biting one of them on the arm. The intruders then abandoned the robbery and fled the scene, thanks to Charlie, another brave parrot.

Chapter 3
Gorilla

For the Greater Good

The older male was the first to arrive at the primitive but deadly snare—a device designed to trap an animal in a rope noose. After a few seconds, a young male and female emerged from the shadows of the underbrush. They joined him at the snare, appeared to consider the situation for a few moments, and then began taking it apart like seasoned pros. They worked quickly and efficiently, without getting caught in the trap themselves. They pulled back the trigger branch and snapped it in half, releasing the tension of the rope noose and rendering the snare inoperable.

Anyone hearing this story would assume the individuals destroying the snare were human, possibly even specially trained soldiers. But they were not. They were wild gorillas in Rwanda who had somehow learned to dismantle poaching snares.

The snares—set to catch prey used as bush meat—are created by making a noose with rope and attaching it to a branch that is pulled downward toward the ground, which creates tension on the rope. The noose is then anchored to the ground with a rock or stick and camouflaged with leaves and branches. When an animal steps on or bumps into the anchoring stick or stone, the branch snaps upward, catching him in the noose. Although the hunters who set the traps claim to have no interest in gorillas, they sometimes catch gorillas nonetheless. Young gorillas can die in the snares, and older gorillas can lose a limb.

Researchers from the Dian Fossey Gorilla Fund's Karisoke Research Centre witnessed the scene described above. They were hiking in Volcanoes National Park in Rwanda when they saw the four-year-old gorillas dismantling the snare. But that was simply the start of the gorillas' "workday." After they finished taking apart the first snare, the same two young gorillas moved on to another snare, where a third juvenile joined them. The trio then dismantled the second snare, working so deftly that it seemed likely that they had done this work before.

It turns out that the day the researchers had witnessed the juvenile gorillas destroying the snares was only a few days after a very young gorilla had been caught in one. The tiny youngster broke his shoulder, presumably while trying to escape, and developed infections that ended his life. Researchers wondered if there might be a connection between the harm the snares had caused the gorillas and their snare-dismantling behavior. Researchers had also previously witnessed older gorillas removing snare wire that was wrapped around the arms of younger gorillas who had escaped from snares, so it seemed likely that the gorillas understood the snares were a threat.

Gorillas are gentle, nonterritorial social primates who appear to be capable of empathy and altruism.

Although this wasn't the first time researchers had seen gorillas destroying poaching snares, it was the first time they had seen such young gorillas doing it, which raised the possibility that the gorillas were teaching their young how to take apart the snares. Extraordinary as it is that gorillas might be passing on this kind of knowledge to their young, even more fascinating is what it suggests about gorilla empathy.

The gorillas who destroyed the snares knew where the traps were, so they didn't need to dismantle them to protect themselves. They simply could have avoided them. By dismantling the snares, they appeared to be protecting *other* gorillas, which allows for the possibility that they were experiencing—and acting on—concern for others. Remarkably, they might be practicing a kind of social welfare.

Nurturing Nature: Binti Jua

Primatologist Craig Demitros knows how important mothering is to young animals, especially to females. If a female infant isn't properly mothered, she often doesn't know how to nurture her own offspring later in life. Lulu, a zoo-born gorilla, had not been taught mothering skills by her own mother. So it wasn't too surprising when Lulu eventually gave birth to a daughter and then neglected her.

Demitros was determined that Lulu's daughter, named Binti Jua (which means "Daughter of Sunlight" in Swahili), would not repeat this cycle of neglect, so he and other zoo staff decided to take on the job of raising Binti Jua, with the express intent of teaching her mothering skills. They hoped—with their nurturing and training—that if she one day had her own offspring, she would know what to do.

Demitros and the zoo staff bottle-fed Binti Jua and socialized her by exposing her to other gorillas. They coaxed Binti Jua to learn nurturing skills using a stuffed animal as a surrogate baby. They taught her how to cradle the stuffed toy and how to "babysit" it, which is a part of maternal care known as "retrieval." To teach this retrieval skill, the zoo staff moved the stuffed toy around, away from Binti Jua, teaching her to retrieve the toy, return it to her area, and keep her "baby" within arm's reach.

Little did Demitros know that Binti Jua would one day use her retrieval skills to rescue a human child. But that is exactly what happened in August 1996, when a three-year-old boy crawled over a fence surrounding Chicago's Brookfield Zoo's gorilla pen and fell twenty-four feet onto the concrete floor of the gorilla enclosure that housed Binti Jua and six other gorillas. At the time, Binti Jua was eight years old and had a seventeen-month-old baby of her own, Koola.

According to zoo staff members and visitors, as soon as Binti Jua noticed the child, she quickly approached him, with Koola riding on her back. She picked up the injured boy and cradled him in her arms, keeping

Gorilla mothers are very protective of their infants and keep them close, by carrying them or letting them ride on their backs.

him away from the other gorillas. She then carried him through the enclosure to the door used by the zookeepers. There, she carefully placed the motionless child on the floor, where the zoo staff were able to retrieve him.

The boy was rushed to the hospital, where he was treated for a broken hand and facial cuts. After four days at the hospital, he made a full recovery. The story and a video of the incident spread throughout the world and turned Binti Jua into an international heroine.

Binti Jua's behavior has been widely interpreted as evidence of gorilla empathy. But there are skeptics who claimed that her nurturing behavior that day was the result of conditioning—the retrieval behavior taught to her by zoo staff—rather than true empathy.

Silverbacks have been known to take a parental role when a young gorilla loses his mother. Under such circumstances, a silverback sometimes steps in and forges a parental bond with the motherless youngster.

But this doesn't seem like a fair dismissal given that even humans respond empathetically to those in need through a combination of innate tendency and learned behaviors. When a human female is taught how to care for a baby—by caring for younger siblings or babysitting for other families—and later becomes a mother herself, it is unlikely anyone would dismiss her mothering behavior as mere "conditioning." It is more likely that good parenting skills—whether demonstrated by a human, gorilla, or other animal—result from a combination of learned behaviors and some sort of innate empathy.

Jambo

Binti Jua isn't the only gorilla who has had to contend with a child falling into her enclosure. In August 1986, while visiting England's Jersey Zoo, five-year-old Levan Merritt fell into the gorilla exhibit. The twelve-foot fall knocked Levan unconscious and fractured his skull. Gorillas began moving toward the injured boy when one of them—a male named Jambo—behaved in a way that forever changed Levan's life. Jambo stood guard over Levan, positioning himself between the boy and other gorillas in what animal behaviorists interpreted as a protective gesture. Jambo even gently stroked Levan's back. When Levan regained consciousness and started to cry, Jambo and the other gorillas retreated, at which point a zookeeper and paramedic removed Levan from the enclosure.

Recorded on video, the incident was reported on major news channels around the world and helped to shift public perception about gorillas. Needless to say, the experience also changed Levan Merritt. Now an adult, with a family of his own, Levan still expresses gratitude to Jambo for his gentle and protective nature.

Frown, Sad, Trouble

Koko was a western lowland gorilla best known for her ability to communicate in sign language. She knew more than one thousand words in American Sign Language and understood roughly two thousand words of spoken English. Koko's ability to communicate with humans enabled us to have a glimpse of a gorilla's inner life. Because she could sign, we learned that Koko was self-aware. When asked what she saw when looking at a mirror, Koko signed "me," and when asked who she was, she answered "Koko" and described herself as a "fine animal gorilla."

We also learned that Koko had a sense of humor. She chuckled at absurdities, such as when someone pretended to feed candy to a toy alligator; she laughed at incongruities, such as when she tied her trainer's shoelaces together and signed "Chase!"; and she wryly called people—and other gorillas—names, such as "Toilet" or "Devil," when she was upset with them or just didn't like them.

But perhaps most important, we learned that Koko had emotions similar to those of humans. She developed friendships, expressed desires (such as wanting

In this photo of Koko as a young gorilla, she is using the sign "listen" to tell her trainer, Penny Patterson, that she would like to listen to the phone.

to have a baby), and cared for a kitten. Unfortunately, the kitten, whom Koko named All Ball, was hit by a car and died. Koko appeared to grieve for her feline companion, signing "Cry," "Frown," "Sad," and "Trouble" when she learned of All Ball's death. She also grieved after learning of the death of actor Robin Williams, whom she had met and befriended when he visited her.

Koko didn't only express sadness after experiencing her own losses, but also seemed to relate to other people's emotions. Whenever Koko watched the film *Tea with Mussolini*, she would turn away during an especially sad part of the story (about a boy who lost his mother) and sign "Frown," "Sad," "Trouble," and "Mother," while her face contorted into expressions her caregivers had learned to associate with sadness.

Koko was a remarkable ape who helped to change the public's perception of gorillas—especially their capacity for communication and empathy.

Chapter 4
Dog

The Crow Pass Guide Dog

In Rudyard Kipling's story "The Cat That Walked by Himself," the first man on Earth awakens to find a dog in his cave. He asks the first woman, "What is Wild Dog doing here?" The woman responds, "His name is not Wild Dog anymore, but First Friend because he will be our friend for always and always and always."

A free-spirited husky named Nanook, better known as "Nookie," is the kind of dog who might have inspired Kipling to write so passionately about canine loyalty. The world came to know about Nookie in June 2018, when Amelia Milling, a college student from Tennessee, was on a three-day solo hike on the Crow Pass Trail in Alaska's Chugach State Park. Late in the afternoon on her first day on the trail, Amelia took a wrong step, went tumbling down a snowy slope, struck a boulder, and flew another few hundred feet down the hill. After landing, she caught her breath and, still lying on the ground, assessed her condition. She was badly banged up but seemed to have no incapacitating injuries. She looked around to see if other hikers might be nearby, but saw no one. As she turned again, she was startled to see what appeared to be a white wolf who seemed to have materialized out of thin air. As Amelia cautiously looked more closely, she spotted something around his neck—a collar and tags—and realized that he was a dog. She also saw that his collar was engraved with the words "Crow Pass Guide."

The mysterious dog's appearance at that challenging moment motivated Amelia to get up, dust herself off, and figure out what to do next. She decided her immediate best course of action was to set up her tent and rest. That's what she did, and eventually, as night fell, she invited her new four-legged friend inside. But the dog made it clear he was a sleep-under-the-stars kind of canine and took his place outside Amelia's tent. As she settled down to sleep, she wondered if he would still be there when she woke up.

In the morning, he was there, waiting for Amelia with his tail wagging. After she packed up, Nookie eagerly led her back to the trail, where together they hiked about seven miles until they arrived at the glacier-fed Eagle River, which they needed to cross to continue on the trail. The dog jumped in and crossed the river, making it look easy. But when Amelia stepped into the water, it was so icy that it took her breath away. As she tentatively took a few more steps, she also discovered that the river was a lot deeper than she expected. Losing her confidence, she backtracked, found what looked like a better place to cross, and entered the river again.

Huskies are known to be intelligent, friendly, and outgoing—traits that likely contributed to Nookie's habit of rescuing hikers on the trail.

The depth of the river at this location seemed shallower and the current slower, but the glacial silt muddied the water so much that Amelia couldn't see the bottom. She didn't make it very far before she slipped and found herself immersed up to her neck in frigid, rushing water. For what felt like fifteen minutes, Amelia struggled, trying to get back to shore, but the current kept pulling her down. She was growing colder by the second and wondered if she would survive. Then Nookie jumped back in the river, swam over to Amelia, grabbed her backpack strap in his mouth, and pulled her out of the water.

Amelia emerged, shivering uncontrollably. She stripped off her wet clothes and crawled into her sleeping bag to warm up. She figured she would rest and then try again later to cross the river. She must have dozed off at some point, because she awoke to the dog licking her face. But something wasn't right; she felt woozy and knew that she was no longer in any shape to try to resume her return journey. She reached for the satellite alert device that she had placed in a waterproof bag in her pocket and, finally admitting to herself that she needed help, pressed the button that sent an SOS signal to Alaska State Troopers. Amelia then curled up in her sleeping bag and fell back into a deep sleep.

According to reports from the state troopers who were in the search helicopter, they eventually spotted a young woman huddled in a red sleeping bag with a white dog lying by her side. They landed, identified

Amelia, and evacuated her to a medical center in Anchorage. They also brought along the dog. Using the address on the dog's collar, they contacted his owner, Scott Swift. Later, while being interviewed by a reporter from the CBC, Scott explained that Nookie was a wild-hearted husky who often vanished for days at a time. It was not unusual for Nookie to be escorted home by hikers or skiers who met him in the backcountry.

Scott had engraved Nookie's collar with the words "Crow Pass Guide Dog" and his home address because of his frequent forays into the wild. The twenty-four-mile Crow Pass Trail head is only a half mile from where Scott lives, and Nookie often followed hikers onto the trail to join them on their hikes. Over the years, two other stories of Nookie helping others had reached Scott. One such story that Scott had heard was about a family of hikers whose eight-year-old daughter had fallen into the Eagle River and was swept downstream. Nookie pulled her out. Then there was the one about a woman who was hiking along a ridge path with friends when she slipped on a rock and started sliding down toward the gorge. Nookie grabbed her jacket and stopped her fall. And now there was Amelia Milling.

Curious about all the stories of Nookie's heroic exploits, Scott wondered if there were other people his wandering husky had rescued or helped out along the trail. So he started a Facebook page. It didn't take long for reports to come in, and Scott was soon shocked by just how many people Nookie had helped out.

As for Amelia, after being released from the local medical center, she spent some time with Scott and Nookie, showering her canine rescuer with dog treats. Not too much later, Nookie was showered with accolades, too. He was named an honorary Alaska State Trooper and an honorary member of the Alaska Solstice Search Dogs, and he was presented the 2019 Red Cross Real Hero Award for Wilderness Rescue.

According to Scott, Nookie still embarks on his solo adventures on the Crow Pass Trail, but now Scott has equipped him with a GPS collar device to make it easier to keep track of him. If you happen to be on the Crow Pass Trail and notice that a white husky has started hiking along with you, his name is probably Nookie. But you can think of him as Kipling might—as First Friend.

Returning the Favor

There's a prevalent feeling among people who adopt shelter dogs that these rescued dogs are somehow more grateful to their humans than other dogs. While there isn't any scientific evidence that this canine gratitude is real, it can be difficult to dissuade some rescue dog families of this belief, especially if their adopted dog behaves in an extraordinary way.

One family who adopted a Doberman pinscher named Khan will always believe that their rescue dog "returned the favor." When they adopted him, the emaciated dog was in bad shape, with broken ribs and other signs of abuse. The shelter had even considered putting him down because of his condition. But Khan's luck finally changed in 2007 when he was adopted by the Svillcic family in Atherton, Australia.

Four days after bringing Khan home, Catherine Svillcic was watching Khan enjoy his new backyard. Her seventeen-month-old daughter, Charlotte, was also playing in the garden. Suddenly, Khan was right there by her toddler's side. Catherine realized that something was amiss when Khan started to behave strangely. He assumed what appeared to be an aggressive posture and started nudging and even tried pushing Charlotte with his snout. When the toddler didn't move, he grabbed her by the diaper and tossed her over his shoulder as if she were a rag doll. Charlotte landed three feet away from Khan just as a snake lunged forward and sunk its fangs into Khan's front paw, causing the dog to yelp in pain.

As Catherine rushed out to her daughter, she understood what had happened: a King Brown snake—the world's third most venomous snake—had been approaching Charlotte. Khan, detecting the snake, had done all he could to move Charlotte out of harm's way, but when he couldn't, he picked her up and tossed her away, taking the bite himself.

Thankfully, Charlotte was just fine. But Khan struggled to make it into the house, where he collapsed.

Catherine immediately brought him to the vet, and after an antivenin shot and several days of observation, Khan made a full recovery. A local snake expert later theorized that Khan had survived the bite of such a venomous snake because it hadn't been able to inject enough venom in his paw. However, if Charlotte had taken that same bite instead, the story most likely would have ended tragically.

Dogs sometimes behave more protectively toward children than adults, possibly because they sense children's vulnerability.

Catherine believes that Khan's actions were deliberate, possibly even an expression of gratitude for having been rescued by the family. Though the idea that Khan was expressing gratitude is speculative, there's no doubt that the recently adopted Doberman put his own life at risk to save the toddler. And for that reason, Catherine pledged that Khan would be loved and well cared for until the end of his days.

Not Just a Warm and Fuzzy Feeling

Most dog owners believe that their canine companions care about their feelings, but scientists have dismissed this belief as nothing more than sentimental projection. However, current research seems to be siding with dog owners. Over the past few years, several studies of dog empathy have demonstrated that dogs respond to the emotions of their human companions, and one recent study reported that dogs who had strong bonds with their owners would even go out of their way to offer help and comfort.

Credit: Belinda Recio

The study, presented in a paper called "Timmy's in the Well: Empathy and Prosocial Helping in Dogs" (named after an episode from the television series *Lassie*), was led by scientist Emily Sanford. According to Sanford, the inspiration for the study was an experience shared by so many dog owners. "Every dog owner has a story about coming home from a long day, sitting down for a cry, and the dog's right there, licking their face," explained Sanford. "In a way, this is the science behind that."

The study involved thirty-four pairs of dogs and owners. Working with one dog-human pair at a time, the researchers asked the owners to sit behind a clear door held shut with magnets and either hum the song "Twinkle, Twinkle Little Star" or pretend to cry. The researchers wanted to find out how the dogs would respond to their owners. Would the dogs try to open the door to reach their owners more often when their owners cried or when they hummed?

Turns out that the number of times the dogs opened the door were about the same regardless of whether their owner was humming or pretending to cry. But the dogs who opened the door when they heard their owner crying opened it three times faster than dogs whose owners were humming.

According to Sanford, the study shows that dogs not only sense what their owners are feeling, but if a dog knows a way to help their owner, they will make an effort to try to help them. "Dogs have been by the side of humans for tens of thousands of years and they've learned to read our social cues," Sanford said. "Dog owners can tell that their dogs sense their feelings. Our findings reinforce that idea, and show that, like Lassie, dogs who know their people are in trouble might spring into action."

A Guardian Angel

On a cold Sunday evening in early January 2010, eleven-year-old Austin Forman, of Boston Bar, British Columbia, went outside to get a load of firewood for his family's wood-burning furnace. His best friend, a golden retriever named Angel, accompanied him. As Austin made his way to the woodshed, he noticed that Angel was staying unusually close to him. Usually, the playful dog would have used the quick trip to the woodshed to explore or run around a bit. But she lingered close to Austin. When he neared the shed, Austin figured out why Angel was behaving strangely—there was some kind of animal in the shadows. At first, Austin thought it was another dog. But as the animal moved into the light, he realized it was a cougar in a crouching position. With only ten feet between them, the cougar leaped toward the boy. In that split second, Angel ran toward them, jumped over a lawn mower, and threw herself between the cougar and Austin.

Within a few more seconds, the cougar had Angel's head in his jaws. Austin ran for his house, shouting for his mother. When Sherri Forman looked out the window and saw what was happening, she immediately called 911. The local constable sped to the scene, and because the cougar was still attacking Angel when he got there, he had no choice but to shoot the cougar. Angel was rushed to the vet, where she was treated for extensive injuries to her head, including surgery for multiple skull fractures. Luckily, with time, Angel recovered.

Austin had always loved Angel, but he told the reporter from television the Canadian Broadcast Network that he feels differently about her since she saved him from the cougar. "She was my best friend, but now she's more than a best friend—she's like my guardian now."

Even though golden retrievers are known more for their friendliness than their protectiveness, many have risked their lives to protect their human family members.

Chapter 5
Lion

When the Lion Lies Down with the Baby Oryx

We admire lions for their beauty, strength, and the regal bearing that earned them the epithet "king of the beasts." When we think of them, we envision majestic predators working together to take down an antelope or other large prey animal, which will be shared later by the pride. What we don't imagine is a lion doting over a baby antelope as if he were her own cub. But such behavior is exactly what made one lioness from Kenya's Samburu National Reserve famous.

Known by the locals as Kamunyak, which means "blessed one" in the Samburu language, the lioness captured the public's imagination in 2002 by "adopting" not one, but at least six baby oryx, a species of antelope. Kamunyak forcibly took each newborn oryx from their mothers and then treated each calf as if he were her own. She would nuzzle the baby, carry him by the scruff of his neck, groom him, protect him from other predators, and act in other maternal ways. She even sacrificed most of her hunting forays, choosing to go hungry in order to watch over her adopted babies.

Despite Kamunyak's efforts to protect her first oryx, another lion ate it when she was sleeping. In response, Kamunyak walked around roaring in what appeared to be a state of distress. After that, she started following herds of oryx. However, Kamunyak never killed any of them despite the fact that lions usually hunt and eat this species. Instead, she hunted warthogs. It seemed that for Kamunyak oryx herds weren't prey, but a source of babies whom she could kidnap and then mother.

It's possible that females who adopt babies belonging to another species do so because they are ready to become mothers themselves, but for one reason or another haven't been able to give birth. Without offspring of their own, they express their maternal instincts toward the young of another species.

After her first oryx was eaten, Kamunyak continued to steal and treat oryx calves as if they were her own cubs. But the next three calves were removed from her custody by park rangers who were afraid the baby oryx would starve to death without milk from their real mothers. They did this despite Kamunyak's truly extraordinary behavior with her third calf: she allowed a female oryx to approach and nurse this third baby for a few minutes each day.

Each time she lost an oryx—whether to another lion or park ranger intervention—Kamunyak once again followed herds of oryx and eventually got herself another baby. When she captured her fifth oryx, rangers decided to take a hands-off approach and let nature take its course. Sadly, this oryx died of hunger, and Kamunyak eventually made a meal of the carcass.

When she ate the last oryx, Kamunyak was behaving more like a normal lion. But her behavior up until the oryx died—as well as her behavior with all the other oryx she had adopted—puzzled scientists. Lion expert Craig Packer theorized that because Kamunyak was a young lioness, closer in development to a teenager than an adult, she was simply "behaving like a youngster." By insisting on mothering the calves, she was acting "like a big kid who's grown too attached to her dolly." He predicted that Kamunyak would stop her oryx adoptions once she had her first litter.

Packer also allowed for the possibility that Kamunyak adopted the oryx because she was suffering from depression, possibly because she was a solitary lioness without a pride. Like Packer, Lawrence Frank—another lion expert—also speculated that Kamunyak's behavior arose out of being alone. In *The Heart of a Lioness,* a documentary about Kamunyak, Frank speculates that if Kamunyak was on her own because she lost her pride, the trauma might have caused her to find comfort and company—rather than nourishment—in the baby oryx.

Scientists had hoped to use a radio collar to track and study Kamunyak, but she disappeared before they had their chance. We don't know if she died of hunger, continued adopting baby oryx, or eventually had her own litter of cubs that she could mother. But what we do know about Kamunyak has taught us that other species—even predators like lions—are capable of a much wider range of behaviors than scientists thought possible.

Rescued by Lions?

If a lion can depart from her normal predatory behavior and mother a baby oryx, is it possible that the king—or queen—of the beasts could show a similar tenderness toward a human? We know that lions raised in captivity are capable of developing close bonds with their human caregivers. Elsa, the lioness made famous by the book *Born Free: A Lioness of Two Worlds* was very affectionate with Joy and George Adamson, the humans who raised her. She nuzzled them, cuddled and slept with them, played with them, and sucked Joy's thumb when she was feeling anxious.

What about lions who are completely wild? Given the inherently predatory nature of these great cats, could they ever look upon a vulnerable human as anything but dinner? This is a question that was raised in 2005 in Addis Ababa, Ethiopia, when lions once again behaved in a distinctly un-leonine way.

In early June of that year, a twelve-year-old girl was walking home from school when seven men kidnapped her, intending to force her into marrying one of them. Tragically, forced marriages still occur in Ethiopia. Some unwanted marriages are arranged

The Leonine Sisterhood

Female lions have a good thing going. They live in matrilineal prides that hunt together, they provide day care for one another's cubs, and they even support equal breeding opportunities with the males. In other carnivorous but social mammal species, one female is usually dominant and therefore the more prolific breeder. But this is not the case in lion society. According to Packard, the "queen of beasts" is unusually democratic, and scientists have seen no sign of breeding dominance among the females. Further, when lions give birth around the same time, they form nursery groups and collectively care for and defend one another's cubs. Finally, as a lioness ages and starts to suffer from physical infirmities that impact her hunting skills, younger lionesses help her to survive by sharing their kills with her. From day care to eldercare, these feminist felines have an admirable sisterhood!

through families, but most of the time they happen through kidnapping.

After beating the girl into submission, the kidnappers continuously moved the girl from location to location in order to avoid capture by the authorities and the girl's family members, who were searching for her. One week later, while still on the run, the kidnappers mistakenly wandered into lion territory. It was then that a group of three lions changed the fate of the schoolgirl by chasing away the kidnappers.

Later that day, police and family members found the girl—and the lions—on the outskirts of Bita Genet, a town about 350 miles away from her home. According to Sergeant Wondimu Wedajo, a member of the law enforcement team that discovered the missing girl, as they approached the scene, they saw three lions and a young girl who was softly sobbing. When the team drew closer, the lions slowly rose, walked away from the girl, and quietly headed back into the forest. Apparently, the lions had stayed with the girl all day without harming her.

In an interview with nbcnews.com, Wedajo credited the lions with the girl's survival. "If the lions had not come to her rescue, then it could have been much worse," he explained. "Often these young girls are raped and severely beaten to force them to accept the marriage." Wedajo also acknowledged that the lions' behavior was very unusual, because normally the lions would have attacked the girl.

Stuart Williams, a wildlife expert with Ethiopia's rural development ministry, speculated that the lions might have spared the girl because she was crying. "A young girl whimpering could be mistaken for the mewing sound from a lion cub, which in turn could explain why they didn't eat her," Williams explained. But lion expert Craig Packard doesn't share Williams's theory. Packard believes that if the lions really were guarding the girl for the better part of that day, it's more likely they were merely biding their time and would have eventually eaten her when they grew hungry.

But whatever the reason for the lions' unusual behavior—whether a rare moment of leonine altruism or full bellies that simply bought the girl time—it changed the fate of a young girl.

Maternal instincts can influence the way species—even predators such as lions—respond to other animals.

Chapter 6
Monkey

Monkey Paramedics

Rhesus macaques—a small Old World monkey species common throughout Asia—are the most wide-ranging, nonhuman primates on the planet. They share about 93 percent of their DNA with humans (chimpanzees, by comparison, share about 98 percent). Macaques also have some cognitive and behavioral similarities to humans. To start with, macaques appear to be self-aware. When presented with a mirror, they look at their reflections and groom themselves, flex muscles, and engage the mirror in ways that suggest they understand the reflection is their own. They are also tool users: they use stones as hammers and anvils to open nuts, shuck oysters, and access other foods. And based on recent observations, it looks like macaques might know some lifesaving techniques, too.

In December 2014, an extraordinary interaction occurred between two rhesus macaques. The incident, which took place at a busy train station in the Indian city of Kanpur, was captured on camera and posted to YouTube. The footage shows a male macaque trying to revive his unconscious companion, who had fallen between the tracks after touching high-tension wires.

The monkey behaved in a way a human paramedic might. For nearly twenty minutes, he performed what appeared to be a repertoire of lifesaving techniques. He lifted his companion's motionless body and repeatedly shook and hit it. When that didn't work, he plunged the body into a trough of muddy water near the tracks. But the water didn't revive his friend either, so he resorted to biting him on his head, face, and neck. (The biting was done without apparent intent to cause injury, as the paramedic monkey didn't sink his canines into his friend's face.) The biting seemed to work: the unconscious monkey, soaked and covered in mud, finally regained consciousness, opened his eyes, and started moving. His rescuer then began grooming him, in what appeared to be an attempt to comfort him.

Macaques hug, hold, and groom one another as a way to deepen their bonds and offer comfort and consolation.

In July 2018, another video of a monkey trying to revive a seemingly dead companion was posted online. Once again in India—this time in the village of Khargone—a monkey had received a shock from a power line. Witnesses videotaped the rescuer monkey as he shook his friend, performed chest compressions, and even tried to breathe into his companion's mouth. But all his attempts failed, and the monkey couldn't be revived.

Why didn't these two "rescuer" monkeys simply walk away, leaving their injured friends and resuming whatever activity they were doing before the accidents occurred? Did they understand that their companions desperately needed help? How did they know what to do?

Scientists have occasionally seen other primates, such as chimpanzees, react to death by shaking the dead body or handling it roughly, as if trying to reanimate it. But these videos show monkeys reacting more precisely. It's impossible to watch the videos and not conclude that the monkeys were trying to revive their companions. This leaves us wondering which is more amazing: monkeys demonstrating empathetic concern and altruistic action or monkeys appearing to understand rudimentary CPR techniques (chest compressions and mouth-to-mouth).

Monkey Midwifery

Like humans, monkeys are socially complex primates. To survive, they need to learn the social hierarchy of their troop and the rules of interaction. Once they understand their ranking and what's expected of them, they will nonetheless do what they can to get ahead, including sneakily breaking the rules by manipulating, deceiving, and exploiting one another in order to achieve their goals.

But monkeys have their soft sides, too, especially the females. Midwifery—the practice of assisting with childbirth—was long considered a uniquely human behavior. Most female animals, especially in the wild, give birth alone, without assistance. But recently scientists have learned that monkeys and bonobo chimpanzees (see chapter 16) also practice midwifery. First, there was a 2013 scientific report of a female black-and-white snub-nosed monkey assisting another female during birth while other monkeys looked on with interest.

It is (Biologically) Better to Give than to Receive

Why would monkeys try to rescue an injured companion or assist a family member or friend during the birthing process? There's no immediate benefit associated with these kinds of altruistic acts. There might be an expectation of future reciprocity—a monkey version of the Golden Rule: "Do unto others as you would have them do unto you." But there's no guarantee that the monkeys who perform good deeds will eventually get their reward. So why risk the time and effort?

If this same question were asked about humans, most people would say that human morality is what motivates us to help others. However, the traits that define *human* morality—such as altruism, compassion, empathy, and reciprocity—also can be seen in nonhuman animals, especially in other primates. So it is possible that morality is more likely to be a broadly biological—rather than solely human—quality.

For example, over fifty years ago, scientists conducted an experiment that looked at altruism in macaques. They discovered that most macaques will consistently choose to go hungry if accessing food results in another monkey getting shocked with an electrical current. Further, the macaques were more likely to engage in this kind of altruistic behavior if they were familiar with the individual who might get shocked.

More recently, a study of Barbary macaque monkeys conducted at London's Roehampton University examined how grooming impacted stress hormone levels. Surprisingly, the results showed that monkeys who were groomed a lot were not the most stress-free. It was the monkeys *who did most of the grooming* who had the lowest stress hormone levels. And this finding isn't unique to Barbary macaque monkeys. In other monkey species, it was also the groomers, as opposed to the groomed, who were most stress-free.

So perhaps biology favors those who give rather than receive. Maybe the monkeys who rescue others, go hungry for others, help another give birth, or readily groom others live less stressful—and potentially longer—lives.

Then, in 2014, Meng Yao, an assistant professor at Peking University, Beijing, and her colleagues witnessed and recorded a much more complex display of midwifery involving two monkeys—this time in a troop of white-headed langur monkeys in China. A pregnant monkey—the younger of the two monkeys—was having contractions while the older "midwife" monkey quietly watched. Once the infant's head and shoulders appeared in the birth canal, the midwife monkey approached and prepared to assist. She stood upright on her hind legs, grasped the emerging baby with both hands, and helped to pull him out of the mother's birth canal. As soon as the baby was out, the midwife monkey held him and licked him. When the mother reached out to indicate she was ready to take her newborn, the

Macaque mothers treat their infants in as much the same way as humans do: they gaze at them lovingly, kiss them, fuss over them, and keep them close.

midwife handed her the baby without hesitation, but remained sitting nearby.

Another report of monkey midwifery occurred in 2016, when researchers witnessed golden snub-nosed monkeys assisting during birth. In actions similar to those of the white-headed langur monkey midwife, the golden snub-nosed monkey helped to pull the baby from the birth canal and held the baby after birth. But the golden snub-nosed monkey exhibited additional midwifery behavior: she groomed the mother during contractions.

Yao suspects that midwifery might be more common among nonhuman primates than previously believed. Most monkeys are born at night, and these wild births are rarely seen by scientists, so there could be many more incidents of monkey midwifery that just haven't been witnessed.

Chapter 7
Dolphin

The Good Citizens of the Sea

Humankind has long had a soft spot for dolphins, no doubt because these sleek, athletic marine mammals are social and playful and they seem to genuinely like people. And then there are the stories—which go back to antiquity—about dolphins' remarkable behaviors: guiding ships to safety, assisting fishermen, playing with children, and saving people from drowning and sharks.

One recent story about these seemingly self-appointed guardians of the sea centers on Ron Howes, a British-born lifeguard living in New Zealand. In October 2004, Ron took his daughter, Niccy, and two of her friends—all training to be lifeguards—on a swim about one hundred yards off a beach in Whangarei, on New Zealand's North Island. Shortly into their swim, Ron and the girls were visited by a pod of dolphins, who began circling them, herding them together, and slapping the water with their tails.

Ron tried to break out of the circle to check out what was going on. Two of the dolphins immediately herded him back to the other swimmers, but not before Ron saw a ten-foot great white shark headed in their direction. According to Ron, the water was crystal clear and the shark was barely two meters away from him. When he saw the shark begin moving toward the girls, his heart practically stopped. That's when the dolphins reacted, too. As Ron later reported to a CBC television interviewer: as the shark closed the distance, the dolphins "went into hyperdrive," creating "a confusion screen around the girls. It was just a mass of fins, backs and . . . human heads."

For roughly forty minutes, the shark lingered nearby, and the dolphins continued to circle the swimmers. It was only when the shark swam away that that the dolphins released the swimmers from their protection. But

Dolphin altruism isn't limited to humans—they have also helped other species, such as, seals, whales, and even dogs.

they still didn't leave—the pod escorted the swimmers back to shore, remaining close until they were safely back on the beach.

After the ordeal, Ron learned that another lifeguard, Matt Fleet, had witnessed the event. Matt was on patrol and had seen the dolphins circling the swimmers and tail-slapping the water. He also confirmed that a shark was loitering nearby.

Scientists don't know why dolphins come to the rescue of humans in trouble. Some believe that dolphins simply have an instinctive response to another mammal in distress. But Diana Reiss, one of the world's foremost dolphin experts, does not believe that their altruistic behavior is purely instinctual: she points out that dolphins are selective about whom and when they help. Because dolphins don't *always* help, Reiss believes that they might be making purposeful decisions rather than acting on instinct.

Another dolphin expert, Denise Herzing, points out that the ways in which dolphins help humans are specific to humans. For example, dolphins once approached Herzing's ship and repeatedly tail-slapped the water just before a big storm moved in. There was no apparent reason for the tail slapping, and Herzing considered the possibility that the dolphins' behavior might have been a warning about the approaching storm. Another time, when Herzing's anchor line broke, dolphins guided her ship to the location of the lost anchor. These kinds of behaviors, which appear to be limited to dolphin-human interactions, support the idea that dolphins adjust their behavior intentionally and are not merely acting on instinct.

Mysterious Minds

In her book *Beautiful Minds*, biologist Maddalena Bearzi describes how as a young girl she was deeply intrigued by stories about dolphins voluntarily helping humans. But by the time she became a scientist, she had become skeptical about the air of enchantment around these tales. Then something happened that forever changed her mind.

There are many stories of dolphins guiding lost swimmers back to shore and leading rescue boats to shipwrecked survivors.

As part of her daily research work, Bearzi, along with a few assistants, was in a small boat following a group of nine bottlenose dolphins in the waters just off Los Angeles. The dolphins were used to her presence and were going about their usual business, lazily foraging for prey. Meanwhile, the researchers were taking photographs of the dolphins' dorsal fins—a technique used to identify individual dolphins—and recording their activities.

Not far from the Malibu pier, the nine dolphins hit the jackpot—a large school of sardines—and formed a tight formation (a typical dolphin technique) that forced the fish together and made it easy to feed on them. But almost immediately after they started to feast on the sardines, one of the dolphins abruptly swam away at a very fast clip. Without skipping a beat, the other dolphins followed suit. Bearzi found the behavior odd, because this group of dolphins usually stayed close to a good find until they had depleted all the fish. Curious, she turned her boat around and followed the dolphins.

Three miles offshore, Bearzi saw the dolphins suddenly stop swimming and form a circle. As her boat drew closer, one of Bearzi's assistants saw a girl with long blonde hair floating in the center of the dolphin circle. Bearzi shouted to her, but the girl only looked at her without responding. Bearzi and her assistants brought the boat closer, and they were alarmed to see that her lips were blue and her face was colorless. They pulled her cold, limp body into their boat, called the coast guard, and headed to shore to get her to a hospital. As they did, Bearzi noticed that all the dolphins had left.

At the hospital, Bearzi learned that the young woman had intended to commit suicide by drowning. Her identification, along with a suicide letter, were sealed in a plastic bag tied around her neck. The doctors told Bearzi that if she hadn't found the young woman when she did, she would have died.

Bearzi still thinks about that day and the dolphins' inexplicable behavior. Why did they leave their sardine meal so suddenly? How did they know there was a human in distress *three miles* farther out to sea? Was this event just a coincidence or something more? Science does not yet have answers to these questions, but one thing is clear: these intelligent marine mammals can behave in mysterious ways that have had profoundly positive impacts on human lives.

They Had His Back

The list of swimmers protected by dolphins grows longer every year.

Long-distance swimmer Adam Walker is grateful to dolphins, too. When he was swimming in New Zealand's Cook Strait in April 2014, he happened to look down into the water beneath and noticed a great white shark trailing him. Walker also saw that a pod of ten dolphins had surrounded him and were swimming with him. The pod remained by his side the entire time the shark was nearby and only left after the shark swam away. "I'd like to think they were protecting me and guiding me home," Adam wrote on his Facebook page. He also jokingly speculated that another reason why the dolphins may have had his back was because the Cook Strait swim supported a nonprofit organization devoted to whale and dolphin conservation.

Kindred Spirits in the Sea

Fascinating as their altruistic behavior is, there's so much more about dolphins that intrigues scientists. To start with, dolphins appear to be self-aware. They recognize their own reflections in a mirror and have signature calls that function as names for themselves and other dolphins. They enjoy playing games, such as creating and manipulating underwater bubble rings, and playing "keep-away" with strands of seaweed. Dolphins also participate in interspecies play with whales, dogs, and humans.

Like humans, dolphins aren't always model citizens. They can be mischievous and sometimes drag birds underwater without any apparent intent to eat them, tease fish by offering them bait and then snatching it away, and play catch with sea turtles—all seemingly just for the fun of it. Dolphins might even "get high" by playing with pufferfish, a species that releases a toxic defensive chemical when they feel threatened. In small doses, the toxin seems to induce a trancelike state that dolphins appear to purposefully induce. And, like most social species, dolphins are capable of aggression. Males are known to bully and gang up on one another, and they will sometimes force themselves on females.

Dolphins are smart. In captivity, they can learn complicated tricks and long strings of commands. They display intelligence in the wild, too: they cooperate with one another, engage in teamwork, and use tools. For example, they use sponges to protect their beaks from rocks and broken coral when foraging. Researchers have even observed mother dolphins teaching the sponge technique to their young, suggesting that dolphins transmit culture from one generation to another. Finally, based on observations of dolphins using a complex echolocation technique, it seems they might be capable of addition, subtraction, multiplication, and ratio comparisons.

Across the Species Divide

Although dogs and dolphins have certain behavioral traits in common, such as playfulness and friendliness toward humans, they haven't shared a common ancestor in more than fifty million years. But this didn't stop two dolphins from coming to the rescue of their distant mammalian cousin.

In 2011, while vacationing in Florida, Audrey D'Alessandro and her husband heard a commotion coming from a nearby canal. Audrey decided to investigate and discovered two dolphins frantically splashing in the canal. Then Audrey noticed that a dog—struggling to stay afloat—was in the water with the dolphins. She immediately called 911 and then jumped into the water herself to try to assist the desperate dog. Luckily, firefighters soon arrived and pulled the dog out of the water. The dog was so exhausted that he couldn't even bark.

The drenched and distressed canine was an eleven-year-old Doberman named Turbo who belonged to a local resident, Cindy Burnett. According to Cindy, Turbo had gone missing the night before, and she and her family had searched everywhere for him—except in the water. By the time Audrey found Turbo in the canal, he had been missing for fifteen hours, and Cindy believes that Turbo might have been in the water the entire time. As she explained, the walls around the canal are high, so once he fell in, Turbo would have been unable to climb out on his own.

But the lucky Doberman survived the ordeal, thanks to the two dolphins who alerted the D'Alessandros to his plight. As fate would have it, when the press showed up to cover the story, the two dolphins made a brief appearance—to take a bow?—and then dove under the water and disappeared.

There are documented friendships between dogs and dolphins. In one such friendship—between a dog named Ben and a dolphin named Duggie—when Ben showed signs of fatigue after they had been swimming together for hours, Duggie helped him to stay afloat by bumping him to the surface with his head. It's possible that the dolphins who were in the water with Turbo provided similar assistance to him.

Chapter 8
Cat

The Order of the Blue Tiger

Most people who live with cats would agree that they have retained more of their wild natures than their canine counterparts. Compared with dogs, our feline companions seem to spend more time on the feral side of the fence. They hunt (often returning from their predatory adventures with unwanted trophy "gifts"); they wander (sometimes far from home and for several days); and they engage in noisy nocturnal turf wars (occasionally returning with battle scars). But despite their untamed spirits, cats often develop very strong bonds with their human companions, and sometimes even put themselves at risk to protect a family member. This is what happened when Tara saw one of her humans in harm's way.

Tara, more formally known as Zatara, is a female tabby cat who lives in California with the Triantafilo family. She simply followed the family home one day when they were walking in a nearby park. As soon as Tara made it clear that she was there to stay, she bonded with the entire family, but developed an especially close relationship with Jeremy, the four-year-old son of Erica and Roger Triantafilo.

One afternoon in May 2014, Jeremy was happily riding his bicycle in his driveway when a neighbor's dog suddenly appeared on the scene. The canine intruder—a Labrador-Chow mix called Scrappy—raced over to Jeremy, lunged at him, and knocked him over. Scrappy then sunk his teeth into Jeremy's leg and shook him back and forth, the way a predator shakes prey. He started to drag Jeremy down the driveway, when out of the blue little Tara leapt into action. The feisty feline pounced on the dog and somehow—despite their size difference—pushed Scrappy off Jeremy. Determined to drive her point home, Tara then chased Scrappy away.

Seconds later, after looking out the window and realizing what was happening to her son, Erica rushed out to tend to Jeremy. She was immediately joined by the triumphant Tara, who appeared by Jeremy's side, as if she wanted to check on her charge, too. Erica rushed Jeremy to the hospital, where he had multiple stitches to address two deep wounds caused by the dog's violent attack.

Although "guard cats" are not as common as guard dogs, there are other stories of cats protecting their owners, including one about a normally friendly tabby cat named Binky. In 2017, when a burglar tied to break into the Indianapolis home that Binky shared with his owner, Binky attacked the intruder and scared him off.

According to the Triantafilos, any time Jeremy—who has autism—was upset or hurt himself, Tara would come running to his side. But the Triantafilos never appreciated Tara's loyalty more than they did the day Scrappy attacked Jeremy. After Tara rescued Jeremy, the bond between the boy and his cat only deepened.

Fortunately, the Triantafilo's surveillance cameras caught Tara's heroics on tape, and they uploaded a video to YouTube to share the story with the world. The video broke YouTube's viewing records at the time by receiving over 16.8 million views in the first forty-eight hours!

As a result of her fame, Tara ended up being honored by numerous organizations and received several awards, including The Military Order of the Blue Tiger Award. This honor, presented by the Exercise Tiger Association, is normally reserved for military service dogs who protect soldiers. Susan Haines, the executive national director of the Exercise Tiger Association, explained "She may be a tabby, but her heart is the heart of a tiger."

The Alarm Cat

In early February 2016, when Ohio resident Dale Lindsey heard the familiar meowing of his "fuzzy alarm clock," he thought Bob was telling him that it was time to get up. Roughly seven years old at the time, Bob was a rescue cat adopted by Dale and his wife, Morena, four years earlier. The Lindseys let Bob sleep on their bed, and over time he developed an uncanny knack for knowing when each of them had to get up. He would start meowing and gently tapping their faces just before their respective alarms went off.

But that day, when Bob woke up Dale, it wasn't just before the alarm clock rang. In fact, it wasn't even

Cats, dogs, and other animals can not only sense chemical changes in the environment but also in the human body. When they alert humans to these changes—by exhibiting specific behaviors—they can call attention to changes in blood sugar, predict seizures, and even help to detect cancer.

morning—it was late on a Saturday night. Dale assumed his cat's internal clock was off, so he tried to go back to sleep. But Bob kept meowing. When Dale continued to ignore him, Bob upped the ante and smacked him with his paw, but not as gently as he usually did to wake him in the morning. When a few emphatic whacks didn't rouse Dale, Bob used the last weapon in his arsenal and bit him. The bite got Dale's attention.

Shocked by the strangely aggressive behavior from his normally gentle cat, Dale chased Bob out of the bedroom. Dale then made his way into the kitchen, thinking that maybe a late-night snack would soothe his unruly feline. But instead of approaching the bowl of food Dale presented to him, Bob went to the door that lead to the attached garage. Thinking that was odd, Dale opened the door and discovered that his car was running inside the closed garage.

The Lindseys had installed a remote starter that day, and it had malfunctioned during the night and started the car. If Bob hadn't awakened Dale, the carbon monoxide from the exhaust could have filled the house and threatened—or even taken—the couples' lives. But thanks to Bob's mysterious behavior, the Lindseys survived to tell the tale.

The Feline First Responder

When an elderly woman in upstate New York discovered an injured three-month-old kitten in a box on her front porch, she asked a local couple, Glen and Brenda Kruger, to help out and take the cat to

Cats in the Coal Mine

There are many stories about cats (and other animals) alerting humans to dangerous levels of carbon monoxide. Animals used deliberately to detect poisonous gases, the way canaries were used in coal mines up until the late twentieth century, are described as "sentinel species" because they provide early warning signs when toxic gases, such as carbon monoxide, are present. In coal mines, caged canaries showed signs of sickness and distress before humans did, an indication that the miners should leave or put on protective respirators.

People who have been saved by their cats and other pets from carbon monoxide report that their pets vocalized loudly, scratched doors, sniffed the air, pawed, swatted, and even bit their humans to get their attention. Scientists aren't certain why animals do this. Their unusual behavior might be because they are feeling unwell themselves or because they are trying to get their humans to pay attention to a change in the environment. In some cases, the cats had access to an exit, but instead of leaving the house, they remained inside and tried to rouse their sleeping family members.

According to a British study, one-third of cat and dog owners say that their pets had, at one time, alerted them to home hazards, such as carbon monoxide leaks and fires. Ten percent of those surveyed believe their pet saved their life or the life of someone they knew.

the local shelter. The Krugers were happy to lend a hand and took the cat to the SPCA. But the kitten didn't stay there very long. The Krugers couldn't stop thinking about her, so they returned the next day and adopted her. Six years passed before they realized what a fateful decision it was to bring the kitten, whom they named Inky, into their family.

Very late on a bitterly cold January night in 2009, Glen Kruger went into his basement to shut down the woodstove for the night. The pull-down trapdoor that led to the attic was located in the ceiling directly above the basement stairs, and it needed repairing. Because of the cold weather, Glen had rigged a temporary fix to get him through the winter, with the intention of repairing it properly in the spring. But after taking care of the woodstove, as Glen climbed back up the basement stairs, the temporary fix failed and the trapdoor came crashing down on his head, knocking him down the stairs. Lying on the basement floor with multiple injuries, he was unable to move. Glen shouted for his wife, Brenda, but she was at the opposite end of the house behind the closed door of their bedroom.

The icy five-degree air from the open attic door was blowing into the basement, adding the risk of hypothermia to Glen's situation. As he was losing hope, Glen noticed Inky's big eyes looking at him from the top of the stairs. In desperation, he shouted to Inky, "Go get Brenda," without any real hope for a response.

Surprisingly, Inky immediately ran off, and Glen later learned that she had gone to the other end of the house and repeatedly slammed herself against the bedroom door, wailing like her life depended on it. The commotion woke Brenda, who followed Inky to the basement, where she discovered her husband. She called emergency services, and Glen was taken by ambulance to the hospital and then transferred to a spine clinic. He had fractured his neck, injured his spine, broken his arm, and dislocated his shoulder. Glen spent the next six months recovering and was left with a permanent disability. But if Inky hadn't alerted Brenda, which resulted in Glen getting the immediate help he needed, he might not be walking—or even alive—today.

Why did Inky respond so appropriately to Glen's request to get Brenda? It's possible that Inky picked up Glen's fear through what scientists call "emotional contagion" and went looking for Brenda for comfort. It's also possible that Inky went to Brenda for direction. Scientists have recently discovered that cats engage in "social referencing," which, like empathy, involves trying to read another's emotions in order to obtain information about a situation. Or perhaps Inky simply wanted to help Glen.

Credit: Kristen A. Kruger

Inky in a field near his home in upstate New York.

Chapter 9
Pig

When Pigs Fly: How a Potbellied Pig Surprised the World

As Andy Warhol once said, everyone gets their fifteen minutes of fame. But one Vietnamese potbellied pig named LuLu got much more than that. Stories about her appeared on the front page of *The New York Times* and in *People, USA Today,* and countless other publications around the world. She even appeared on television: *Good Morning America,* the *Regis & Kathie Lee Show,* the *Oprah Winfrey Show, National Geographic, Animal Planet,* the *Late Show with David Letterman,* and others.

How did a potbellied pig attract so much attention? It started in 1997, when Jo Ann and Jack Altsman of Beaver Falls, Pennsylvania, agreed to babysit the now renowned potbellied pig. At the time, LuLu weighed only four pounds and belonged to their daughter, Jackie, who just needed her mom and dad to pitch in with a little pig care for a few days. But a few days turned into weeks, then months, and finally the Altsmans realized that LuLu wasn't going home to Jackie. LuLu was theirs, regardless of the fact that, one year later, she had grown into a 150-pound adolescent pig. But Jo Ann and Jack had already made room for LuLu in their hearts, so they found a way to make room for her in their home—and their vacation trailer.

In August 1998, while vacationing in Presque Isle, Pennsylvania, Jack was out fishing and Jo Ann was relaxing in their trailer with LuLu and their dog, Bear. Suddenly, Jo Ann felt a familiar and terrifying pain in her chest. She had suffered a heart attack eighteen months earlier and knew she was having another. Jo Ann reached for the phone, only to fall to the floor. Trying to get the attention of passersby, she yelled for help. She even threw an alarm clock through one of the trailer windows. But no one responded.

Potbellied pigs are intelligent and affectionate social animals who can develop strong bonds with humans and other species.

for an average dog, not a 150-pound pig), getting badly scraped in the process. A few minutes later she returned and appeared to check on Jo Ann, sniffing and nuzzling her. After this tender inspection, LuLu left again. Over the next forty-five minutes, the frazzled pig continued to return, check on Jo Ann, and leave.

Finally, Jo Ann heard a man yelling, "Lady, your pig's in distress!" Jo Ann shouted back that *she* was in distress, too, and told him to call an ambulance. Jo Ann was transported by helicopter to a hospital where she underwent emergency open-heart surgery. The doctors told her that if she if she had gone unattended for much longer, she would have died.

Luckily, Jo Ann made a full recovery. She later found out—from witnesses—that, after busting through the dog door, LuLu had pushed open the gate in the fence enclosing the trailer, ran to the road, and laid down in the middle of traffic, playing dead. LuLu loved to "play dead" with Jo Ann and Jack, probably because it was a reliable way to get their attention. But this time, instead of getting anyone's attention, people ignored LuLu and simply drove around her. After playing dead for a few minutes, LuLu would run back to Jo Ann, sniff and nuzzle her, then run back to the road to repeat her "dead piggy" routine.

Eventually, it worked: someone finally stopped. A man driving on the road noticed LuLu's injuries from squeezing through the dog door and pulled over to

Meanwhile, Bear was barking nonstop, possibly because he sensed Jo Ann's stress. Unfortunately, most people don't find barking dogs a cause for alarm, so Bear didn't attract any attention. However, LuLu responded differently. She softly whimpered while nuzzling and licking Jo Ann, who was lying on the floor with tears streaming down her face.

Then, out of the blue, LuLu left Jo Ann's side and forced herself through the trailer's dog door (sized help because it looked like she needed medical attention. The kind stranger followed LuLu back to the trailer to tell someone that the pig needed help. That's when he discovered Jo Ann on the floor.

Like Jo Ann, LuLu made a full recovery. But life wasn't the same for a long time because of all the media attention. LuLu quickly became a world-famous pig. People loved the story of her heroic actions, and they were amazed that a pig could behave so intelligently and empathetically. For a long time, pigs didn't have the best reputation with people who didn't know much about them, which is why the word *pig* has so often been used in a pejorative sense. But farmers and scientists have long known just how remarkable pigs are. They consider them to be the smartest of all domesticated animals, including dogs. Even when compared to all animals, many rankings of animal intelligence place pigs among the top ten smartest animals on Earth, and a few even place them in the top five.

Pulling Her Weight

When Prudence couldn't reach the gate to open it, she moved bales of hay, climbed on them, released the latch, and opened the gate . . . with her snout. Prudence, better known as "Pru," was a pig who lived on a farm in western Wales with her owner, Dee Jones. Before Pru learned to open the gate, Dee had already figured out that Pru was smart—she was able to quickly learn commands such as "Sit," "Lie down," "Come here," and "Go home." Pru had also demonstrated that she was capable of helping out on the farm—bringing in the cattle and sheep as if she were one of the herding dogs. But it was Pru's self-taught gate trick that convinced Dee she had an especially clever pig, so she decided to save Pru from the slaughterhouse and keep her as a pet.

Many pigs, given the chance, demonstrate that they can herd sheep as well as any dog.

One day, when Pru was only four months old, Dee was enjoying a walk in the country with Pru and her sheepdog when she took a wrong turn and fell into a bog. Getting stuck in a bog can be very dangerous: the waterlogged mosses—hidden under floating carpets of vegetation—can suck a person into the soggy earth, and it is difficult to get out without help. So, when Dee suddenly found herself waist-deep in a thick, muddy morass, she panicked. Then she had an idea. She still had the dog leash in her hand, so she called her loyal pig close enough to loop the leash around her ample girth and commanded Pru, "Go home! Go home!"

"If You Were a Pig, You Would Have This Figured Out by Now"

Charles Darwin once said, "I have observed great sagacity in swine." Tina Widowski, an animal welfare scientist who studies pig behavior, agrees with Darwin's assessment of pig intelligence. To make her point about porcine IQ, Widowski likes to tell people that when she was working with monkeys—known for their cleverness—she would look at them and say, "If you were a pig, you would have this figured out by now."

Once you learn even just a little bit about pigs, you can see why Darwin, Widowski, and many other scientists feel this way. Pigs are remarkably intelligent, sensitive, and self-aware animals. They also appear to understand that others have minds of their own, which in comparative psychology is called having a "theory of mind." Like dogs, pigs respond to their names and can learn the names of other animals and objects. Pigs can be taught tricks, such as "fetch the ball" or "jump over the dumbbell." As for their own communication system, pigs have more than twenty different sounds for different situations, ranging from vocalizations piglets use to call for their mothers to those that alert other pigs to danger.

Scientists recently learned that pigs are tool users. They taught pigs both how to use a mirror to find the location of a hidden object and how to use a joystick (with their snout) to move an image on a computer screen onto a target in exchange for a treat. Pigs excelled at both tasks. These barnyard brainiacs are even capable of learning to adjust a thermostat in a barn. After scientists demonstrated the purpose of the thermostat, the pigs would turn it up when they were cold and down when they were too warm. And, as Pru's gate trick demonstrates, pigs also are consummate escape artists. There are countless stories of pigs figuring out how to unlock and untie nearly every kind of gate closure.

The pig's emotional life is no less impressive than his intellect. To start with, pigs develop close bonds with one another and sometimes appear to grieve the loss of mates, family members, and companions. They also are empathetic and respond to the emotional states of other pigs. People who share their lives with pigs will tell you that pig empathy extends to humans, too. For example, in the book *The Good, Good Pig*, author Sy Montgomery describes her pig Christopher as behaving more gently and with greater patience with shy children.

Dee believed that Pru sensed her distress and immediately cooperated. As Pru started walking toward home, her forward motion—combined with her weight and strength—pulled Dee onto solid ground. When later interviewed by the BBC, Dee emphatically opined, "Without Pru, I wouldn't have been able to get out of the mire." Pru more than pulled her own weight that day, further convincing Dee that she was lucky to share her life with such a smart pig.

Chapter 10
Bear

The Benevolent Bear

Else Poulsen was a bear whisperer. A biologist who devoted her life to the care and study of bears, she spent many years with them in zoos and wildlife sanctuaries. At Canada's Calgary Zoo, Poulsen got to know an especially memorable pair of grizzly bears who were housed together—Louise and Khutzy. Poulsen described Louise as a complex, intelligent bear who quickly came to understand humans by learning their words, gestures, and postures. In her book *Smiling Bears: A Zookeeper Explores the Behavior and Emotional Life of Bears,* Poulsen describes Louise as "listening" to conversations about veterinary procedures scheduled for the following day and then hiding the next morning so that she was nowhere in sight when the vet showed up.

Louise was housed with Khutzy, a cub who had lost her mother. Louise quickly assumed a maternal role and treated the young cub as her own. She overindulged Khutzy, who grew into a spoiled and often rebellious young bear and developed several challenging behaviors that would drive any zookeeper crazy. But Poulsen eventually discovered a management strategy that kept the young bear in line: she conspired with Louise.

The zoo had a safety policy requiring all animals to move into indoor enclosures overnight, but the bears didn't care much for the practice, as it limited their time outdoors. Nonetheless, Louise usually complied with the request to move inside, even if she sometimes required encouragement with food. Khutzy, on the other hand, refused to cooperate if she wasn't in the mood. On these occasions, Poulson had to repeatedly coax the insubordinate bear indoors by offering her treats, catch-me games, and other bribes.

One afternoon, Khutzy decided to make things more challenging by vanishing. Poulsen scanned every corner of the enclosure but couldn't see Khutzy anywhere. Exasperated, she jokingly asked Louise, "Where's

Khutzy?" The last thing Poulsen expected was an answer. But sure enough, Louise "answered" her by turning around and staring at a large boulder. It turned out that Khutzy was hiding behind it. Poulsen then coaxed the reluctant young bear inside with the usual bribes. This routine continued over many nights, with Khutzy hiding and Louise pointing out Khutzy's whereabouts to Poulsen, who eventually lured her indoors.

Recent research suggests that bears are as intelligent as higher primates.

Then one evening, Khutzy absolutely refused to go inside. In a moment of exasperation, Poulsen turned to Louise and jokingly said, "Go get Khutzy." Once again, Louise surprised Poulsen. The bear sighed and wearily made her way across the enclosure to where Khutzy was hiding. Louise herded Khutzy indoors by nipping at her heels. From that day forward, Poulsen relied on Louise to get Khutzy indoors for the night, deeply appreciative of what she described as Louise's ongoing "acts of benevolence."

Kinder than the Average Bear

Do bears, like people, have personalities? Are some bears gentler than others? The zookeepers at Hungary's Budapest Zoo thought so. Two brown bears—a female, Vali, and a male, Defoe—were once housed together at the zoo. As the zoo staff got to know the bears, who were living together in the same enclosure, they formed opinions about the bears' personalities, and it didn't take long for them to declare that Vali had the kinder disposition. And one day, later made famous by a viral video, Vali did something that seemed to prove them right.

At the zoo, other animals—such as wild crows—occasionally make their way into the bears' enclosure. Like bears, crows are clever and resourceful omnivores who will eat almost anything they can find or steal. In 2014, when local crows discovered the bounty at the Budapest Zoo, they made the most of it, stealing meat from lions and tigers, and even attacking smaller residents, such as guinea pigs. The first few crows who found the all-you-can-eat buffet at the zoo must have spread the word, because the zoo staff quickly realized that they had a crow infestation problem.

One day during this crow inundation, while visitors were watching and videotaping Vali and Defoe in

Captive animals often show greater tolerance, and even friendliness, toward other species.

him by grabbing hold of his wing with her mouth. Seemingly startled, the crow pecked at Vali's snout, and she immediately dropped him onto the ground. At that point, Vali either lost interest or felt her work was done, and she ambled back over to her stash of apples and carrots to resume eating.

It took a few seconds for the dazed crow to catch his breath and regain his composure. Then he flipped onto his feet and just stood there, staring at the bear. As for Vali, the video of her "crow rescue" went viral and raised questions about her motivation. Was she helping the crow because the struggling bird appeared to need assistance? Or was she tired of eating carrots and apples and thought the crow might make a good snack, but the peck on the snout dissuaded her? The zoo staff suggested that perhaps Vali was simply curious about the crow. But they also said that the crow was lucky it was Vali who had scooped him out of the pool, because had it been Defoe—who is not as gentle—the story might not have ended as well.

Guardians of the Forest?

When a three-year-old North Carolina boy, Casey Hathaway, went missing in the forest for three days in late January 2019, search and rescue teams from across the state worked tirelessly to find the toddler before it was too late. After a stressful search, they finally found Casey, who was suffering from a low body temperature and mild frostbite. When asked how he survived on his own in the woods, Casey said that a bear had kept him company.

their enclosure, one of the marauding crows injured himself and ended up in the bears' pool. Vali, who was nearby snacking on apples and carrots, noticed the crow flailing about in the water and flapping his wings as if struggling to stay afloat. Vali appeared curious and approached the crow by climbing onto a rock at one end of the pool. Apparently, she wanted to get even closer, so she moved to the other end of the pool where she could reach the struggling bird. Vali then gently scooped the crow out of the water. When the crow started to slip out of her paw, Vali caught

Of course, no one knows if a bear was actually with Casey. Children often deal with fear by telling themselves comforting stories. There are similar accounts of other children lost in the woods later claiming that they were with a bear or other wild animal. In 1955, two-year-old Ida Mae Curtis went missing in

Bear expert Ben Kilham believes that bears know how to read a human's body language. If a bear did encounter Casey, and understood that the child wasn't a threat based on his body language, perhaps she lingered near him out of curiosity, giving Casey the impression that a bear "kept him company."

the Kootenai National Forest in Montana. After two days of searching for her in pouring rain, rescuers found Ida, who said that a bear had comforted her during the night. Earlier still, in 1880, toddler Katie Flynn disappeared in Michigan's Manistee National Forest. Hunters and trappers searched the woods for two days until they finally found Katie, who said she had been guarded by a bear.

Given that many children sleep with teddy bears that comfort them, it's likely that these children's reports are simply soothing stories they had told themselves to help them survive a frightening experience. However, black bear expert Ben Kilham has observed altruistic behavior in bears. As a researcher who works in the field, Kilham has been protected by female bears who kept aggressive male bears away from him. He has also witnessed female bears acting in protective ways toward other bears' cubs. This kind of behavior allows for the possibility that in certain situations, bears, especially female bears, might behave in protective, nurturing ways toward a nonthreatening child lost in the woods.

Bear in Mind: They're a Lot Like Us

- Bears smile. Like humans, the sides of their mouth move upward when they feel good—when relaxing, meeting a friend, playing, or watching other bears play.

- Bears help each other out. Under the right circumstances, bears will share food and even keep an eye on one another's cubs.

- Bears make tools. They fashion back-scratchers from branches and throw rocks, snowballs, and other objects.

- Bears "count." They can distinguish between larger and smaller quantities, which helps them to find the best foraging areas—and to know how many Oreos™ come in a package. Black bear expert Ben Kilham has witnessed the cookie-counting ability of black bears many times. He had a routine in which he would offer Squirty, a black bear he was studying, a sleeve of a dozen Oreos, doling them out one by one as treats. After Squirty had consumed all twelve, she would walk away. Then one day, Ben decided to withhold a couple of cookies by removing them from the sleeve before meeting up with Squirty. When he did this, Squirty seemed to know how many there were supposed to be in a sleeve, and she made it clear that she wanted the missing cookies by looking for them. Ben repeated this "counting test" several times, always with the same results—if he didn't hand over all twelve cookies, Squirty would look for the one or two Ben had removed. So we know that at least one black bear—Squirty—could "count" to twelve.

Chapter 11
Horse

A Horse as Good as Kerry Gold

When we think of dairy cows, we picture them as calm, docile animals, and for the most part they are. But like any animal, under certain conditions cows can become aggressive. One of these conditions is motherhood. The drive to protect her calf can turn an otherwise gentle cow into a fiercely protective and dangerous animal.

It was this kind of bovine maternal instinct that resulted in Scottish dairy farmer Fiona Boyd getting trapped underneath a raging cow weighing over half a ton. Fiona and her husband, Matt, run two dairy farms. Fiona began that fateful day by responding to a call from Matt, who asked if she could move a cow and her calf out of the paddock and separate them. It's a common practice on dairy farms to separate calves from mothers shortly after birth. Fiona usually accomplished this task by walking both mother and calf to the calf house, leaving the calf there, and then bringing the mother to the milking parlor.

However, on this occasion, things did not go as planned. Fiona could only get the two-day-old calf to follow her. She was expecting the mother to catch up with them, but she didn't. What's more, as she was leading the calf, the other cows in the paddock became curious and gathered around them. The commotion caused by the curious cows distressed the young calf, and she began to call out. The mother heard her calf's distress and responded immediately by charging Fiona and knocking her to the ground.

Fiona knew that it's never a good idea to be on the ground among cows. In a situation like that—when a mother is acting in a fiercely protective way and other cows are picking up on her agitation—they will sometimes group together and attack, occasionally with fatal results. She knew she had to get up, but every time she tried, the cow butted her back to the ground. Fiona shouted and fought back, but she was unable to get away from the massive cow who seemed determined to crush her. Finally realizing that there was nothing she could do, Fiona curled up into a ball and shielded her head, accepting the possibility that she might die. But just as Fiona gave up hope, the raging cow suddenly stopped attacking. Fiona's chestnut mare, Kerry Gold, had appeared on the scene.

New research suggests that horses have a basic understanding of human emotions, which might play a role in their ability to forge bonds with humans.

While the attack was unfolding, Kerry had been grazing in the same paddock. At some point, Kerry must have noticed what was going on and decided to take action by charging onto the scene and repeatedly kicking the cow with her rear hooves. This forced the cow to retreat, giving Fiona the chance to crawl under the fence to safety and call for help.

Fiona went to a local hospital, was treated for cuts and bruises, and made a complete recovery. There was no doubt in Fiona's mind that she might not be here today to tell her story if Kerry hadn't intervened. As for the heroic horse, Kerry Gold received extra apples and carrots, as well as endless gratitude from Fiona. Interestingly, after Fiona's attack, Kerry became a self-appointed bodyguard. She would behave very protectively not only of Fiona, but of just about anyone who entered her field.

Rescue Reciprocity

When the sheriff's office in a southwestern Louisiana town received a call that an elderly and extremely emaciated mare was seen wandering in the wild and eating tree bark, they came to her rescue. Soon after getting her sheltered, they enlisted the help of a local organization called SOAR (Steeds of Acceptance & Renewal), who agreed to take on the job of restoring the old mare's health.

As the mare's condition improved, the volunteers at SOAR realized that she was well-bred and had been properly trained at some point in her life. It wasn't long before the mare, whom they named Stormy, regained her health enough to demonstrate that she could still wear a saddle. Once Stormy was able to carry a rider, SOAR decided to find the best home they could for her. They heard that a local family— the Leonards—were searching for a horse for their young daughter, and they determined that the family was a good match for Stormy.

Straight from the Horse's Heart

Many people who have close bonds with horses will tell you that their relationship is based on mutual empathy—on both human and horse being able to read each other's emotional states and recognize the nuanced expressions of those states. Recent research conducted by scientists at England's Universities of Sussex and Portsmouth suggests that this empathetic bond isn't just a sentimental feeling shared by equine enthusiasts—horses may indeed be able to discern and even remember human emotions.

Researchers designed an experiment using photographs of humans with either an angry or a happy expression on their faces. They showed these photos to twenty-one horses and recorded their responses. When shown the photographs of the angry faces, the horses' heart rates increased, and they showed a preference in how they looked at the images: the horses turned their heads so that they could look at the photo with their left eye, which sends information to the right brain hemisphere. This is not surprising, as horses—and other species—are known to process possible threats with the right half of their brain. When shown the photographs of happy faces, the horses showed the opposite bias, preferring to look with their right eye, thereby sending information to the left brain hemisphere, which processes positive social stimuli.

A few hours after showing the horses the photos, the people who had been in the photographs appeared before the horses in person, with neutral expressions on their faces. The horses' only exposure to these people were through the photos they had briefly viewed earlier. To further control the experiment and prevent unconscious cueing, the models did not know which of their photos—the happy face or angry face—the horses had been shown.

Surprisingly, the horses appeared to remember each person and their previous mood from the photographs. The horses who had seen the angry face photograph warily checked out that person with their left eye, and their heart rates increased, just as they had when viewing the photos. The horses who had seen the happy face photo did not show a left- or right-gaze bias, but their heart rates remained normal.

The study provides evidence that horses can read human facial expressions and even remember a person's previous mood when encountering them later in the day, both of which support the idea that horses can empathize with humans. It's possible that equine empathy played a role in motivating Kerry Gold and Stormy to protect their human companions.

The Leonards' daughter, Emma, was nine years old when they adopted Stormy, and her younger brother, Liam, was seven. Emma spent time with Stormy every day, but one day became particularly memorable. In September 2010, on an otherwise ordinary afternoon, Emma decided to take Stormy for a ride, with Liam following along on foot. When Emma saw a new trail, she and her brother decided to explore it. But by the time they reached the end of the trail, something wasn't right with Stormy. She was snorting and acting spooked, which wasn't like her. Emma did her best to calm Stormy, but she couldn't. As Emma was trying to figure out what to do, both she and her brother heard a sound coming from the woods behind them. Emma turned and saw what had been getting Stormy so worked up—a very large wild boar with formidable tusks.

Emma and her brother froze in fear. As the boar advanced and positioned himself between the siblings, Emma prepared for the worst. She had heard stories about territorial wild boars, which were common in the area, attacking people. But then Stormy went into action and stepped in front of the boar. Using her muzzle, she nudged Liam into the nearby woods. She then turned back to face down the boar, which held his ground. Emma hung on as Stormy quickly spun around and kicked the boar in the mouth with her rear hooves. The boar squealed in pain and bolted for the woods. Thanks to the brave old mare they had welcomed into their lives, Emma and Liam were safe.

Stallions protect mares and their herds, and mares protect their foals, so it's likely that protective instincts also play a role in how horses respond to their human companions.

Chapter 12
Seal and Sea Lion

A Protective Circle of Seals

Long ago, in the British Isles, people told stories about shape-shifting seals known as selkies. These supernatural beings were seals when in the sea, but they assumed a human form when on land. Folktales described selkies engaging in all kinds of interactions with humans, from seducing unsuspecting men and women to helping fishermen with their catch and keeping swimmers safe.

If Charlene Camburn's encounter with seals had happened long ago, people might have believed that selkies were involved. In 1999, Charlene, along with her boyfriend, Chris, and seven-year-old son, Brogan, were visiting Donna Nook, a nature preserve on the eastern coast of England, on the North Sea. The preserve is famous for its gray seals, and Charlene and her family had wandered across the intertidal sand dunes to a sandbar to watch the seals. As the late afternoon light started to fade, they began walking back, but the tide had come in faster than expected, partially submerging the sandbar and cutting it off far from the shore. Neither Chris nor Brogan were swimmers, so Charlene decided to strip down to her underwear, dive into the icy waters of the North Sea, and swim the distance to shore for help.

The current was strong and the water was cold, but Charlene swam steadily. Yet after only ten minutes, she felt herself tiring and becoming disoriented. It was only then that she noticed that she was surrounded by a group of six seals. At first, she thought they were just checking her out and would soon be on their way. But the seals lingered—one in front, and the rest in a circle around her. As the sun set, Charlene became increasingly exhausted and chilled to the bone. She started to feel it would be easier to give up and accept death rather than continue to swim. Numerous times she started to sink under the surface, but every time the seals dove down and gently nudged her feet to push her back up. "I could feel the seals," she later told reporters from

The Mirror, a London newspaper. "They were nudging my legs and feet and kept diving beside me and I kept bobbing back up."

When a harbor seal wants to check out something going on at the surface, he rises by "spyhopping"—rapidly moving his rear flippers back and forth, similar to how a human treads water.

Charlene found the seals' presence calming and comforting. "Without those seals around me, I would have given up," she told reporters. "All I could see were their faces around me. Not one of them faced away and they were barking and squealing." One seal in particular was especially reassuring. "I will never forget the seal in front of me," Charlene explained. "He was there all the time, swimming backwards and staring at me. It was a dreamy feeling."

At some point, Charlene saw what she thought was a boat. She tried shouting for help, but her voice was weak. When the boat vanished—along with her

hope—she again started to sink underwater. The seals were immediately on top of the situation, nudging Charlene back to the surface. No one knows how much time passed, but at some point Charlene hazily remembered the seals lifting her yet again to the surface and she saw the bright lights of a boat, realizing that someone had finally come to her rescue. Then she passed out. She had been in the icy waters of the North Sea for nearly two hours.

When she first woke in a hospital bed, Charlene's first thought was about Chris and Brogan. She thought they must have been overtaken by the rising tide and drowned. Thankfully, the reality was that they had found another sandbar that connected to the shore and had run as fast as they could to get help. When the Coast Guard boat arrived on the scene, one of the rescuers fired a flare to light the sea and they spotted Charlene—encircled by seals. After pulling her out of the frigid sea, they determined she was suffering from severe hypothermia and quickly transported her to a nearby hospital. According to a Coast Guard spokesman, if Charlene had remained in the water just a few minutes longer, she wouldn't have survived to tell her story.

Charlene was relieved and incredibly grateful that Chris and Brogan had found a way off the sandbar and had called for help. But she was also grateful to the seals, as was Chris, who was convinced that they had saved Charlene. "Without the seals creating a circle around her, she would have drifted three to four miles further out," he told reporters. Charlene felt the same way and fervently believed that the seals had kept her alive.

The seals had stayed with Charlene nearly the entire two hours she was in the water. They surrounded her, watched her, and vocalized. But they only touched her when she started to sink. If the seals had considered Charlene a threat, they probably would have behaved defensively. Seals can be territorial and, when feeling threatened, have been known to bite humans. Alternatively, if the seals were simply curious about her, presumably they would have checked her out and then went on their way. But they neither attacked Charlene nor inspected her and swam away. They stayed with her the entire time and repeatedly pushed her to the surface every time she started to sink. Why?

There are two species of seal native to Donna Nook: Harbor (also called Common) seals and Gray seals. Both species are known to be gregarious at certain times—such as during their pupping, breeding, and molting seasons. It's possible that Charlene encountered the seals during their pupping season when they are more sociable, and, sensing her distress, they responded as if she were one of their own.

An Unlikely Lifeguard

In northern England, the River Tees flows eastward until it reaches the North Sea near the town of Middlesbrough. Harbor seals once made their home in the nearby estuary known as Seal Sands, but they left the area after it became industrialized. Recently, however, the seals have been returning to the estuary, and sightings of seals lounging on the shore or popping up out of the water are becoming increasingly common. But what local resident Chris Hinds saw in June 2002 was anything but ordinary.

Chris and his son, Raymond, were walking along the river with their dogs when they saw what appeared to be a stray dog behaving oddly. The dog was injured—he had a visible cut on his head—and he was whimpering as he stumbled along the water's edge. As Chris drew closer, he could see that the dog was older and was struggling. Suddenly, the dog—most likely in a moment of confusion—jumped into the river, and the fast-flowing tides began to carry him away. Raymond ran off to call the fire department for help, while Chris stayed on the bank of the river, watching the dog so he could direct the rescuers to his whereabouts.

As Chris watched, the dog started to sink under the surface and cry out. At that critical moment, a seal suddenly appeared. His head popped out of the water, and he started to circle the distressed dog. Chris was stunned by what he saw next. Using his snout, the seal persistently pushed the dog toward the shore until he was out of the water and on the mudflats. "I just couldn't believe what was happening in front of my eyes, it was a truly amazing experience I will never forget," he told reporters from *The Telegraph*.

Seals and dogs occasionally seem curious about each other and will sometimes even interact in a friendly and playful way when meeting on the shore.

The firemen arrived just after the dog was out of the water, and they started to treat his wounds. Meanwhile, two other seals had joined the first seal, and the three of them watched the firemen tend to the injured dog. "By the time we arrived, the dog was on dry land and there were three seals bobbing in the water keeping an eye on him," Mark Baxter, of the Stockton fire station, reported.

When asked about the unusual incident, Dominic Waddell, an expert in seal behavior at Scarborough's Sealife and Marine Sanctuary, suggested that the harbor seal, a member of a naturally protective and nonaggressive species, might have checked out the dog, sensed that he didn't belong in the water, and then pushed him to the shore. Waddell said that the seal's behavior might be similar to that exhibited by dolphins when they protect swimmers from sharks. Whatever the reason, the injured dog owed his life to his unlikely lifeguard.

A Sea Lion's Gift of a Second Chance

In 2000, Kevin Hines, a teenager struggling with bipolar disorder, decided to end his suffering by jumping from San Francisco's Golden Gate Bridge. Miraculously, he survived the fall. But while in the water, he saw what he thought was a shark circling directly beneath him. Kevin braced himself for the worst, wondering why he was spared in the fall only to die in the jaws of a shark. But the shark did not attack him. Instead, he kept circling, and as Kevin began to sink, the shark actually came to him and bumped him back up to the surface. The shark remained nearby, and it wasn't until a rescue boat arrived on the scene that the mysterious animal swam away.

After a lengthy recovery from back surgery due to an injury from the fall, Kevin appeared on a television show about suicide prevention and told his story. Later, a man who saw the program wrote to Kevin to explain that he had been on the bridge that day, had witnessed his jump, and had seen the sea creature interacting with him. But the man offered a detail that until then Kevin had been unaware of. The man wrote, "It haunted me until this day; it was no shark, it was a sea lion, and the people above looking down believed it to be keeping you afloat . . ."

Kevin later learned that others who had witnessed the incident thought they saw a different animal: some of the observers were convinced the animal was a seal, not a sea lion. And a few even insisted that they didn't see any animal in the water with him.

Although we can't know for sure what happened, here is what is likely and why. The animal probably wasn't a shark, as a shark would have almost certainly attacked, not aided, Kevin. It may indeed have been a sea lion, as there is a population of sea lions who live in the San Francisco Bay area. Finally, sea lions are known to be social and inquisitive animals, and there are many reports and videos of sea lions circling, investigating, and even engaging in playful ways with swimmers and divers who pass through their territory. So, the most likely explanation is that it *was* a sea lion who rescued Kevin.

The story has a happy ending, against many odds, as the Golden Gate Bridge is the world's top spot for

Sea lions are curious about humans and often approach divers and playfully interact with them.

suicide attempts. Ninety-eight percent of all jumpers die. Kevin is one of the rare 2 percent who survive. He eventually became a mental health advocate and delivers suicide-prevention talks around the world. And, not surprisingly, his website features a silhouette of a sea lion behind his name, an expression of gratitude for the animal he credits with saving his life.

A Case for Sea Lion Empathy?

Sea lions are known for their intelligence, playfulness, and sociability. They often stay very close together, piling up on one another as they lounge on rocks or float in "rafts" with dozens of other sea lions. These inquisitive sea mammals often check out swimmers and divers who enter their territory, and there are many stories throughout history about positive encounters between sea lions and humans. There is foot-age on YouTube of curious sea lions checking out and playing with divers and engaging with aquarium visitors through the tank's glass walls, and even one video of a sea lion pup jumping onto a boat and snuggling—like a puppy—with the boat owner for more than an hour.

Renowned ethologist Marc Bekoff believes that sea lions have empathetic natures based on his observations of sea lion mothers. While witnessing their pups being attacked and eaten by orcas, Bekoff reported that they "squeal eerily and wail pitifully, lamenting their loss."

A recent study that looked at yawning in sea lions allows for the possibility that sea lions engage in "contagious yawning," which can be an indication of empathy. In its simplest form, empathy is a physical response to another's movement or mood, such as "catching" a yawn when another person yawns. If it turns out that sea lions do engage in contagious yawning, it would be further evidence of their empathy.

Finally, there's the sea lion named Ronan, who was the first nonhuman mammal to demonstrate the ability to find and keep a beat. Ronan became famous for his ability to dance to the Backstreet Boys' song, "Everybody." If you're wondering what this has to do with empathy, studies show that highly empathetic people are more receptive and responsive to music. Perhaps it's the same for sea lions?

Chapter 13
Orangutan

The Heart and Mind of an Orangutan

Chantek wasn't an ordinary orangutan. He knew how to communicate with humans using more than 150 gestural signs based on American Sign Language. He also understood many words of spoken English. Born in 1977, Chantek was raised by humans in a human setting as part of a language research project directed by anthropologist Lyn Miles. As a result of this study, this extraordinary ape gave the scientific community its first in-depth look at an orangutan mind.

With a vocabulary similar to that of a human toddler, Chantek used signs to request favorite foods, rides to restaurants, cage cleaning, and more. He demonstrated linguistic creativity by coining his own terms when he didn't know the sign for something, such as signing "tomato" and "toothpaste" when referring to ketchup. Chantek made and used tools, including those required for arts and crafts, such as painting and jewelry-making. He understood the concept of money, and in exchange for doing chores, he received an allowance that he liked to spend on fast food. Chantek also had excellent navigational skills, and when taken out for a drive by Miles or research assistants he would often direct the driver to the local Dairy Queen®, where he would "purchase" a cheeseburger.

But Chantek wasn't merely clever. He also demonstrated a depth of emotion. Chantek developed close relationships with Miles and other caregivers and often used signs to ask to see them and to express concern about them. For example, if in the course of imaginative play his caregivers pretended that a toy animal was on the "attack," Chantek would "protect" the caregivers from the attacking toy. He also could be quite chivalrous. Once, when Chantek and Miles were outside and it started to rain, he picked up a scrap of fabric, tore it in half, and handed one half to Miles. He held the other half above his head as if it were an umbrella, looked at Miles, and signed "rain." Another time, Miles had a scratch on her hand, and Chantek asked her about it. After Miles explained that she had hurt herself cleaning, Chantek asked to touch and kiss her hand.

Orangutans are intelligent primates who are capable of observational learning. After quietly watching humans engage in their daily activities, wild orangutans have performed such tasks as paddling a canoe, cutting wood with a handsaw, washing themselves with soap, and much more—all without any instruction.

Later in his life, when Chantek was relocated to Zoo Atlanta to be part of another cognitive research project, he worked with a research assistant named Hannah Jaicks. The young college graduate was taken with Chantek's gentle nature, and she often spent extra time signing with him to get to know him better. One day Jaicks arrived late for work, in a frazzled state of mind. She had recently cut herself, and her hand was wrapped in a large, cumbersome bandage. While she raced to get her research equipment set up, she ignored Chantek, who was persistently signing to her. Finally unable to ignore him any longer, she

impatiently asked him, "What, what is it you want to tell me?" Chantek responded by signing that he felt sad. Still frustrated, Jaicks replied, "Sad? Chantek, why, why are you sad? We need to work." Chantek signed, "I am sad because you're hurt." Chantek's empathy touched Jaicks deeply, so she set aside the day's research goals and spent the rest of the day just visiting with him, enjoying his incredible heart and mind.

The Friendly Orange Apes

Orangutans have a reputation for being nice guys, as demonstrated by the following stories about three captive males. The first story involves an orangutan once housed at the Dublin Zoo in Ireland. In 2011, a videotape was uploaded to YouTube showing Jorong's rescue of a drowning moorhen chick (a type of waterbird). The footage shows the gentle ape checking out the listless bird and first offering her a leaf. When she doesn't grab hold of the leaf, he lifts her out of the pond and places her on the ground, where he gently strokes and sniffs her. A spokeswoman for Dublin Zoo said that this incident didn't surprise her because Jorong had a reputation for being sweet and inquisitive.

In 2015, at the Phoenix Zoo in Miyazaki, Japan, an orangutan named Happy was caught on camera sharing his food with chimpanzees housed in a cage across the aisle. Uploaded to YouTube, the video shows Happy reaching through the bars of his enclosure and gently tossing food to chimpanzees who are

Primatologist Leif Cocks describes orangutans as gentle, peaceful animals who are known to forge lifelong friendships with one another and their human caregivers, too.

begging with open hands. Orangutans and chimpanzees do not live in the same parts of the world, but that didn't stop Happy from sharing his food with his simian cousins.

Hsing Hsing, a male orangutan who was housed at Australia's Perth Zoo until his death in 2017, was another orange ape known for having a sweet nature. To start with, he would share his food with a female orangutan named Utama—she wasn't his mate, just a good friend. He also would cooperate in the most trusting way with his caregivers. He had diabetes and learned to calmly present his arm for daily insulin injections and blood glucose monitoring. But the incident that best reflects Hsing Hsing's generosity of heart was when he escaped his enclosure in 2016. Keepers were able to catch him quickly because rather than speedily making his way out of the zoo, he stopped to try to free his orangutan friends.

The Person of the Forest

Orangutans—the only great apes found outside of Africa—live in rainforests on the islands of Borneo and Sumatra. The name *orangutan* means "person of the forest"—an apt name for an ape who shares 97 percent of our DNA. Primatologist Leif Cocks describes orangutans as "the most peaceful of all on our family tree." Cocks and other scientists who study orangutans find these critically endangered great apes to be patient, compassionate, and intelligent. Although orangutans will defend their young to their *own* death, they do not attack to kill the way chimpanzees and humans do.

When it comes to cognitive abilities, orangutans are no less intelligent than African apes. In fact, their capacity for paying attention, thinking insightfully, and manipulating objects might even exceed that of other apes. Perhaps this is why zookeepers consider them to be the most ingenious escape artists of all the great apes. Zoologist Ben Beck once described the difference between ape species by comparing how they would respond if given a screwdriver: A chimpanzee would try to use the tool for everything except its intended purpose; a gorilla would first be afraid of it, then try to eat it, and then forget about it; but an orangutan would "notice the tool at once but ignore it lest a keeper discover the oversight. If a keeper did notice, the ape would rush to the tool and surrender it only in trade for a quantity of preferred food. If the keeper did not notice, the ape would wait until night and then proceed to use the screwdriver to pick the locks or dismantle the cage and escape."

In addition to their amazing minds, orangutans are known for being extraordinarily devoted mothers. They carry and breast-feed their young until they are six years old, and they continue to tend to them—teaching them about which foods to eat, where to find these foods, and how to build a sleeping nest—until they are ten years old. In fact, young female orangutans will sometimes linger even longer with their mothers—into their teens—in order to learn parenting skills by observing their mothers raise younger siblings. The only other animal with such a prolonged relationship between mother and offspring is humans.

Chapter 14
Elephant

An Elephant Never Forgets

When trainer Darrick Thomson makes his rounds at the Elephant Nature Park sanctuary in Thailand, he gets a very special greeting from Kham La, a young female elephant. She enthusiastically trots toward Thomson the second she sees him. She inspects him with her trunk and softly trumpets the deep, rumbling sound that elephants make when they feel calm. Thomson responds by murmuring endearments and stroking her, which, based on the way she leans into Thomson, appears to be just what the young elephant wants. From these affectionate greetings, as well as all the other interactions between the two, it's clear to everyone at the Elephant Nature Park that Thomson and Kham La have forged a special bond, which began the day Thomson rescued Kham La.

Kham La was born in captivity at an elephant trekking camp—a business that offers elephant rides to tourists. Despite the seemingly benign purpose of the camp, the methods employed to train elephants to carry tourists are often cruel. Shortly after birth, Kham La was separated from her mother, who was a working elephant in the camp. Despite being just a baby, Kham La was immediately subjected to the inhumane training technique known as "crushing" or "spirit-breaking," which involves beating a very young elephant until she submits to human domination.

Luckily, fate intervened for both Kham La and her mother when, in 2015, Thomson rescued them and brought them to the sanctuary. At that time, Kham La was four years old, which is still quite young for an elephant, so the staff tried to reunite her with her mother. Sadly, the older elephant was too traumatized to forge a new bond with Kham La. But over time, other, older females at the rescue sanctuary became nannies to Kham La, and she eventually became a member of a herd.

From the moment Kham La arrived at the sanctuary, Thomson worked with her, showering her with affection and gaining her trust. Kham La responded in kind, and soon all Thomson had to do was call her name and she would run to him like a puppy. Today, Kham La is so close to Thomson that she even holds his hand with her trunk when they walk or run together, or when they are just sitting quietly next to each other.

One especially endearing exchange between this unlikely duo was captured on video and posted online. The footage shows Thomson, wading into the river at the sanctuary, while Kham La and her adopted herd are quietly hanging out on the opposite bank. As Thomson swims, he splashes up a fair amount of water, which catches Kham La's attention. Apparently, she must think he is in trouble, because she quickly makes her way into the river toward him. By the time she reaches Thomson, he is almost back on the riverbank, but Kham La nonetheless tries to lift him up out of the water with her trunk in what looks like an attempt to "save him."

Even though Thomson wasn't actually in danger, it appeared to all who witnessed the attempted "rescue" that Kham La believed he was. Her altruistic response may have been motivated by the bond she shares with Thomson. After all, none of the other elephants rushed to his assistance. Given what we know about elephantine memory, it's very likely Kham La remembered that Thomson had saved her from the camp. Those memories, coupled with his kind behavior toward her at the sanctuary, may mean she will be devoted to him for the rest of her long life.

The Kindness of Elephants

Elephants are known for their kindness and compassion: they care for and protect one another by providing assistance to ill, injured, and trapped herd members. They have been seen holding one another up when injured, rescuing one another's calves from river currents and predators, and removing darts and spears from one another's bodies. And as we learned from the story about Kham La and Darrick Thomson, we know that elephants are capable of forming strong bonds with human caretakers and acting altruistically toward them. But it seems elephant empathy extends to other species, as well.

Having a good memory is critical to an elephant's survival. It helps them to remember and distinguish between friends and foes, and to recall where the best feeding grounds and watering holes are located.

The Amazing Elephant

Greek philosopher Aristotle described the elephant as "the animal that surpasses all others in wit and mind." Today, most scientists would agree: they consider the elephant to be one of the most intelligent of all species. Elephants engage in tool use, such as using branches for accessing food that is out of reach of their trunks, and they cooperate with one another to solve problems, such as working together to rescue a calf. They recognize themselves in mirrors, which is one of the criteria scientists use to determine if an animal is self-aware. Their impressive memories enable them to remember one another, as well as other animals and humans, over several decades. Elephants communicate over long distances (as far as six miles) by making very low rumbling sounds that they "listen" to with the pads on their feet. Their incredibly sensitive feet can also sense approaching storms that are as far as 150 miles away. In captivity, elephants have demonstrated that they can mimic both human speech and the vocalizations of other elephant species. Finally, elephants perform ritual-like behaviors around newly dead elephants and even over the bones of long-dead elephants, suggesting that they have an understanding of death.

In his book *The Beauty of the Beasts*, animal behaviorist Ralph Helfer tells such a story about an elephant and rhinoceroses that took place in East Africa. He was on a safari and had stopped to watch a mother rhino and her baby as they approached one of several salt licks left out by ranchers. After indulging in a long lick, the mother rhino started to make her way back to the forest. As the baby rhino followed her, he got stuck in the mud. There had been three days of torrential rains, and the earth had softened into several feet of mud. The young rhino cried out in distress, but his mother did not rescue him. She backtracked, walked around him and sniffed him, but then inexplicably made her way back into the forest, leaving her calf behind.

Soon after, a herd of elephants approached the salt lick near the baby rhino. The youngster once again cried out. His mother came running out of the forest and charged the elephants, who decided to move on to another salt lick about one hundred feet away. Satisfied that her baby was safe, the mother rhino returned to the edge of the forest to forage.

A few minutes later, an adult elephant from the newly arrived herd approached the stuck baby rhino and gently inspected him with his trunk. What happened next completely surprised Helfer. The massive elephant, who had an impressive set of tusks, gently knelt next to the young rhino and carefully placed his tusks underneath him, using them as a forklift in what looked like an attempt to raise the baby rhino out of the mud. But before he finished, the mother rhino reappeared from the brush and charged him again. The elephant released the baby, stepped aside for a few moments, and returned to his herd.

Over the next few hours, this interaction played itself out repeatedly: the adult elephant appeared to try to rescue the baby rhino, but the mother rhino stopped him by charging. Finally, the elephant gave up, and he, along with his herd, moved on. Luckily, later in the day, the baby rhino freed himself and rejoined his mother.

Helfer speculated that the mother rhino simply didn't have the intelligence to figure out how to help her calf. Luckily, the elephant did. Helfer considers the elephant's repeated attempts to help the baby rhino as a standout example, among the many he has personally witnessed, of elephant altruism.

Elephants are considered to be one of the most empathetic animals and have shown compassion toward not only one another but also members of other species.

Chapter 15

Hippopotamus

The River Crossing Guards

Every year, as part of their migratory route, 1.5 million gazelle, wildebeest (also known as gnus), and zebras cross the Mara River, which flows through Kenya and Tanzania. Known as the Great Migration, their journey is the largest land migration on Earth, with many animals traveling more than five hundred miles during each migratory cycle. It's not an easy journey for these animals—the river currents are strong, and crocodiles lurk in the shallows. Over 250,000 animals die every year from predators, hunger, thirst, and exhaustion. But despite the risks, the animals make the journey in order to reach pastures and drinking water.

Tourists travel from far and wide to watch this breathtaking spectacle, which is considered to be one of the most dramatic on Earth. But in October 2010, in addition to bearing witness to millions of animals risking their lives to complete their migration, tourists in Tanzania got a bonus. As they watched countless animals enter the raging river, they saw a young wildebeest get swept downstream, away from her mother. Given the strength of the current, this wasn't unusual. But what they saw next was: a female hippopotamus came to the little wildebeest's rescue.

Tom Yule, who runs Grumeti Expeditions tour company, was at the river, watching the hippopotamus's extraordinary behavior and shooting video. Yule reported in a talk radio interview, "The hippo immediately went after it and positioned itself on the downstream side of the calf, nudging it with its snout and keeping it above the water all the way across the river until the calf reached the other side." Abdul Karim, an experienced safari guide, also witnessed the rescue. "To everyone's amazement the hippo came to the gnu's rescue and pushed it gently to the river bank," Karim explained in an online news report.

After the calf was safely on shore, the young wildebeest caught his breath and rejoined the herd. Then, less than ten minutes later, a young zebra was seen struggling with the current, and the same hippopotamus once again took action, pushing her across the river to the opposite bank. "Maternal love can be so strong it can even surpass species boundaries," Karim theorized in the same report.

Hippopotamus mothers and their young will often form a group with other cows and calves for protection against predators.

Hippopotamuses are very protective of their own calves and have often been seen chasing crocodiles away from their young and helping them to cross rivers. But the rescues described here involved the young of other species. Furthermore, this kind of interspecies rescue by hippopotamuses wasn't observed just this once. There are other stories of hippopotamuses saving animals in trouble. In 2014, when visiting the Masai Mara Reserve in Kenya, Israeli photographer Vadim Onishchenko captured footage of a hippopotamus coming to the rescue of a wildebeest in a river under attack by a crocodile. His video shows a group of hippopotamuses taking notice of the wildebeest being attacked. One hippo then steps forward, scares off the crocodile, and proceeds to persistently nudge the seemingly confused and exhausted wildebeest toward the riverbank, while guarding him from further crocodile attacks.

Ironically, hippopotamuses have the reputation of being the most dangerous animal in Africa. These semiaquatic giants might seem harmless as they lazily submerge themselves in rivers, lakes, and swamps, but they are responsible for more human deaths—an estimated five hundred per year—than any other African animal. The hippopotamus' territorial and protective nature can result in aggressive and unpredictable behavior. But now that hippopotamuses have also demonstrated altruistic behaviors, it's clear that there's much more to these animals than we realized.

Hippopotamus Empathy

The ancient Greeks named the hippopotamus after the horse (in fact, the word *hippopotamus* translates to "river horse") because the basic shape of the animal was horse-like. Modern scientists, in contrast, might have named the hippopotamus *river pig* because they believed the hippopotamus was most closely related to pigs. But recent advances in molecular phylogenetics, which analyzes genetic material at the molecular level, primarily in DNA, shows that the hippopotamuses' closest relatives are cetaceans—whales, dolphins, and porpoises.

The idea that whales and hippopotamuses are related is interesting for many reasons, but one reason in particular stands out. The hippopotamus brain has spindle neurons, which are special cells associated with emotions, empathy, and social interactions. Whales, chimpanzees, elephants, and humans also have spindle cells. Since species with spindle cells engage in altruistic behavior, perhaps these cells are playing a role in the hippopotamuses' interspecies rescues.

Scientists recently learned of another unexpected hippopotamus behavior that supports the idea that they experience empathy. When an animal cares for a sick, injured, or dead member of its own species, those actions are referred to as epimeletic behavior—and hippopotamuses have been seen engaging in these kinds of activities.

In Botswana's Chobe National Park, in September 2018, scientist Victoria Inman was the first person to observe and report on epimeletic behavior in hippopotamuses. Inman watched a female hippopotamus tenderly caring for the carcass of a roughly six-month-old calf, which was floating in the lagoon. Inman assumed the female was the calf's mother based on her swollen mammary glands and protective behavior. The presumptive mother tried to keep the dead calf afloat. She also guarded it, chased crocodiles away from it, and even blew bubbles around the carcass, which is one of the ways hippopotamuses communicate with one another.

Later in the day, other members of the hippopotamus pod moved from the river to the lagoon, where they interacted with the corpse much in the same way the mother had—they swam with it and kept it afloat. Were these hippopotamuses trying to help the calf? Were they feeling grief?

Chapter 16
Chimpanzee and Bonobo

Please Person Hug: Washoe's Generosity of Heart

Washoe enjoyed playing with toys, looking at picture books, painting pictures, hosting tea parties, and checking out people's shoes. In addition to playing, Washoe understood that she was expected to follow certain rules, such as brushing her teeth and doing her chores. The structure of her vocal cords made it difficult for her to speak, so Washoe communicated using sign language, which she used to request things she wanted, such as asking to see her visitors' footwear. By the end of her life, Washoe had learned roughly 350 American Sign Language signs.

Washoe might sound like a speech-impaired child, but she was a chimpanzee raised by Allen and Beatrix Gardner in the 1960s as part of a language learning experiment. Washoe has the distinction of being the first nonhuman to learn to communicate using American Sign Language. The Gardners immersed Washoe in human culture and treated her as if she were a human child: Washoe wore clothes, sat a table, slept in a bed with sheets and blankets, used the toilet, rode in a car, and participated in other activities typical of young children.

When Washoe was five, she was moved to the University of Oklahoma's Institute of Primate Studies in Norman, Oklahoma, where Roger and Deborah Fouts took over her care. In his book *Next of Kin: My Conversations with Chimpanzees*, Roger Fouts tells numerous stories about Washoe. Fouts paints a fascinating picture of Washoe's intelligence, but his stories about her empathetic and compassionate nature are what is most compelling.

Like humans, chimpanzees will hug, hold hands, pat backs, and kiss to console one another and to demonstrate affection.

One especially poignant story involved a volunteer named Kat. In 1982, Kat was pregnant, and Washoe was very interested in her expanding belly and gestating baby. Sadly, Kat miscarried and didn't go to work for several days. Her absence upset Washoe, so when Kat returned to work, Washoe was pouty and gave her the cold shoulder. Kat decided to tell Washoe why she had been away. Washoe had lost two of her own infants—one baby had died of a heart defect shortly after birth and the other of a staph infection at two months—so Kat thought Washoe might more readily forgive her if Washoe knew the reason for her absence. Kat signed, "My baby died," and Washoe responded by lowering her eyes and sitting quietly. She then looked up, met Kat's eyes, and signed "Cry" by placing a finger under her eye and drawing it down over her cheek, like a falling tear. Kat was deeply moved by

Washoe's expression of empathy. Shortly afterward, Washoe signed again, this time "Please person hug," further demonstrating that she felt Kat's loss and wanted to console her.

Washoe behaved similarly after Fouts broke his arm and showed up at the Institute in a sling. Upon seeing Fouts, Washoe signed "There," pointed at Fouts's arm, and approached the wire-mesh partition that surrounded her enclosure. Washoe pushed her fingers through the mesh, gently stroked Fouts's injured arm, and softly vocalized in what clearly seemed like an attempt to console him.

It wasn't just humans who elicited empathy from Washoe—she behaved compassionately to other chimpanzees as well. One recipient of Washoe's kindness was Bruno, another chimpanzee at the Institute of Primate Studies. Bruno had been raised by humans since birth and consequently lacked basic chimpanzee survival skills, including a fear of snakes. Oklahoma is home to dozens of species of snakes, and one day several snakes found their way into the chimpanzees' outdoor island enclosure. One of the other chimpanzees saw the snakes and let out a warning call. Washoe and all the other chimps—except for Bruno—responded appropriately and moved away from the snakes to the other end of the island. But then Washoe noticed that Bruno had remained where he was, seemingly unaware of the nearby danger. Washoe immediately signed "Come hug! Come hug!" to coax Bruno toward her, but Bruno hadn't learned any sign language yet, so he didn't respond. Without hesitating, Washoe backtracked to the other end of the island, grabbed Bruno's hand, and led him to safety.

As remarkable as these expressions of empathy are, the event that made the greatest impression on Fouts involved another chimpanzee, named Penny. Shortly after Penny arrived at the Institute, she climbed over a fence and fell into the moat that surrounded the chimpanzees' island. In response, Washoe behaved like a superhero. She got a running start and jumped over the fence and into the moat. Given that chimpanzees can't swim and have a fear of water, this was truly extraordinary behavior. Washoe then held on to a fence post on the edge of the moat with one hand, reached out to Penny with the other hand, and pulled the terrified newcomer to safety. Fouts could not believe the degree of heroism Washoe had demonstrated that day. After all, Washoe had only known Penny a few hours and yet still risked her life to save a virtual stranger.

What Is It Like to Be a Bonobo Thinking About a Starling?

In 1974, philosopher Thomas Nagel wrote an essay entitled "What Is It Like to Be a Bat?," in which he argued that because consciousness is subjective and varies a great deal from one species to another, we can never really know what it's like to be another species. He used bats to make his point, because bats experience the world so very differently from humans. For example, bats fly, hang upside down, and use echolocation, whereas humans do none of these things. Although we can *intellectually* understand what they are experiencing, we can never actually *know* what it's like to be a bat. That said, we try to imagine how other species experience the world. But what about animals? Do they understand the differences between themselves and other animals? Are they capable of imagining what it's like to be another species?

In his book *Our Inner Ape*, Frans de Waal tells a story about a bonobo named Kuni who was housed at England's Twycross Zoo. When Kuni saw a starling fly into a pane of glass and fall to the ground, she immediately reacted by attempting to get the starling airborne. First, she helped the stunned bird to stand, and then she gently tossed the bird into the air. When that didn't work, she climbed a tree with the bird, carefully unfolded his wings, and launched him like a paper airplane. The starling still hadn't recovered enough to fly, and he once again fell to the ground. At this point, Kuni climbed down and watched over the starling for several hours until he was fully recovered and flew away.

Given that apes demonstrate empathy, it's possible they are capable of imagining. Scientists have observed wild chimpanzees playing with small logs as if these objects were infants. The chimpanzees cuddle, cradle, and put their log babies to bed in nests that they make for them. Captive apes, including bonobos, have behaved similarly with dolls and stuffed toys.

What Kuni displayed in her attempts to help the starling was a complex level of perspective-taking and empathy. She had to think about what it's like to be a creature entirely different from herself, recognize that the creature needed help, and imagine the kind of help the creature needed (in this case, resuming flight). Not only was Kuni capable of doing all of this, but her motivation for doing so appeared to be altruistic, as there was no material reward or benefit for Kuni in trying to help the starling.

The Sisterhood of Motherhood: Bonobo Midwifery

Every woman who has either given birth, assisted at a birth, or witnessed a birth knows that it isn't easy bringing a baby into the world. For this reason, from at least the beginning of written history, mothers-to-be were assisted by midwives—women trained to assist in childbirth. There are references to midwifery in

Kindness of the Apes

Chimpanzees and bonobos are our closest living relatives, sharing approximately 99 percent of their genome with us. They also appear to share certain emotional capacities with us: they are capable of generosity, kindness, empathy, and compassion. Over the past decade, researchers around the world have designed numerous experiments that reveal just how charitable humankind's closest relatives can be. Here are some examples of chimpanzee and bonobo kindness:

• Bonobos will share food with strangers (other unfamiliar bonobos) and help a stranger get food, even if they don't get food or any other benefit for themselves in return.

• Chimpanzees will provide helpful assistance, such as handing another chimp an object that is out of his reach. They even appear to think about *how* to be helpful: they will carefully watch what another chimp is doing and then hand them a tool that would be specifically useful to the task, even if the other chimp isn't looking for it.

• Both bonobos and chimpanzees will comfort friends who are in distress. After conflicts, they comfort the individual who was on the losing end of a skirmish by grooming, hugging, kissing, stroking, and playing with him, which seems to lower the defeated chimp's stress level.

• Chimpanzees, like many humans, treat others with greater kindness and compassion when they themselves have been the recipients of such gestures.

ancient Greek and Roman texts, as well as in the Bible. Until recently, humans were thought to be the only species to provide assistance during birth. The accepted theory was that midwifery developed because human births are difficult, due to women's narrow pelvises and infants' large heads. But now scientists are considering the possibility that midwifery developed for other reasons as well, because it's not practiced only by humans.

Bonobos—also known as pygmy chimpanzees—are a separate species of chimpanzee characterized by a slightly smaller build, female dominance, and the use of sex for conflict resolution. In 2018, scientists witnessed three separate occasions when female bonobos, who were living in primate parks in The Netherlands and France, assisted pregnant females with their deliveries. And they acted in the same kinds of ways that humans do when assisting at births.

The female bonobo "midwives" tended to be mothers, and therefore had already had experience with birthing. When they realized that labor had begun, they gathered around the pregnant female, protected her from males, swatted away any insects, and frequently sniffed at the birth fluids, which might have provided them with information about how far along the delivery was. They also positioned themselves at the entrance to the birth canal with cupped hands in an apparent attempt to catch the baby as he emerged.

Although discovering midwifery behavior in animals was surprising to scientists, finding it in bonobos made sense to them because female bonobos have strong bonds, and females, rather than males, often dominate their troops. Chimpanzees, in contrast, are patriarchal, and the females give birth alone. Researchers think it's possible that midwifery might have evolved in a common human-chimp-bonobo ancestor, but over time, chimps lost the trait.

Female bonobos don't just assist one another with birth—they defend one another against male aggression. After a male bonobo attacks a female, a group of females will gather together and form a coalition to punish the offending male. Older females appear to be in charge of these coalitions, which mostly protect younger females who are more vulnerable to attacks.

Chapter 17
Crow

Under His Wing

One quiet day in 1999, Ann and Wally Collito, from Massachusetts, noticed something furry scurrying around the edge of their property. It was small, so they thought it might be a rat, but they soon realized it was a young kitten and wondered if someone had abandoned it. A few days later, they saw a crow hanging around the kitten. The Colitos worried that the crow might be hunting her, but they quickly realized that the crow wasn't stalking the kitten—he was "babysitting" her!

The Collitos named the kitten Cassie and called her feathered guardian Moses. They watched in disbelief as Moses groomed Cassie, fed her worms and insects, and kept her safely off the nearby busy street by cawing at her when she started to wander toward traffic. Certain that no one would believe what they were seeing, they decided to videotape Moses and Cassie so that they could share the story with the world.

Over time, the couple coaxed Cassie into their home, where they began to feed and care for her. But that didn't end her relationship with Moses. Every morning, he would call on Cassie by pecking at the Collitos's door. They obliged by letting Cassie out, and the unlikely pals played for hours at a time—wrestling, exploring, and lounging around together. This continued for five years, until one day when Moses didn't peck at the door. Given that crows usually live for only seven to eight years in the wild, the Collitos assumed Moses had died.

Although we do not know what Moses was thinking or feeling when he was caring for Cassie, it seems that he somehow sensed she needed help, which allows for the possibility that Moses felt empathy for Cassie. This kind of empathy—between two completely unrelated species—is especially rare. Empathy is not hard to imagine among animals belonging to the same species or among different species who live together, such

The Feathered Apes

Crows (and other corvids) have often been called "feathered apes" because of their intelligence and emotions. These charismatic and clever birds "count" by distinguishing between quantities. They make and use tools. They have a "theory of mind," meaning they understand that others have their own separate minds and thoughts. They play and cooperate with one another, form alliances, and share resources. They have a sense of fairness, engage in reciprocity, and bring gifts to people who feed them or help them recover from injuries. They appear to have an understanding of death and may even hold simple "funerals."

as dogs and people, because in these cases empathy benefits both parties. When members of the same species are sensitive to and help one another, it is in the interest of the species as a whole; and when dogs comfort their human companions, they are likely to receive affection and food in return. But when a crow cares for a kitten, the only benefit to the crow is friendship. That's always been enough to inspire empathetic behavior in humans, so perhaps it's sometimes enough for other animals, too.

Chapter 18
Rat and Mouse

Rats to the Rescue? Rethinking the Much-Maligned Rodent

Charles Darwin once said that animals whose communities "included the greatest number of the most sympathetic members would flourish best, and rear the greatest number of offspring." Despite being prolific breeders, it's hard to imagine that Darwin was thinking about rats when he said this, considering all the expressions that paint these rodents as anything but sympathetic. For example, when someone cheats another, we might call them "a dirty rat." Snitching on someone is "ratting on them," and when we sense betrayal, we say that we "smell a rat." Finally, we describe the meaningless pursuit of wealth and power as a "rat race."

Do these negative expressions reflect the true nature of rats? Researchers Inbal Ben-Ami Bartal and Peggy Mason had a hunch that rats don't deserve their bad reputation, so they designed an experiment to test rats for empathy. They placed a pair of rats in a clear acrylic box, confining one to a clear plastic tube with a door at one end that could only be opened from the outside. The other rat was free to roam. Rats tend to avoid open spaces and usually prefer to stay in corners or travel along walls. But in nearly every test, the free-roaming rat left the safety of her corner, figured out how to open the door, and released the trapped rat. Once free, friendly nuzzling often followed.

Bartal and Mason believe that the free rat sensed the trapped rat's distress, became distressed herself, and tried to help. The researchers were impressed with the empathy the rats demonstrated, but they wondered how far rat empathy would go. So, they brought chocolate—which rats love—into the equation. This time the free rat had two choices: open the door to release the trapped rat or open the door to a different tube that contained five chocolate chips.

Rats are highly social animals who are known to play, sleep, forage, and nest together. According to a recent study, rats are so social that they can even read the emotional cues expressed in a fellow rat's face. When a rat feels pain, his facial expression changes, and other rats appear to understand that the expression signals distress.

These experiments are just a few of many that have tested empathy in rats. All of the tests had similar results: the free rats spared trapped rats from many kinds of unpleasant and even painful situations. One of the first tests, conducted in 1959, put two rats in a divided cage. The rat in one side of the cage could get food by pressing a lever, but the same lever delivered an electric shock to the rat in the other side of the cage. Time and again, as soon as the rat with access to the lever realized the consequences, he stopped pressing the lever and went hungry. In a similar kind of test, one rat was put in water and the other was placed next to it in a dry enclosure with a door. Ninety percent of the time, the dry rat saved his buddy from drowning by opening the door. Even when tempted with chocolate, the majority of the time, the dry rat saved the swimming rat first.

Most of the time, the free rat first chose to release the trapped rat and then go for the chocolate. These results might suggest that, for rats, helping their cage mate is on a par with chocolate. The free rat could take advantage of his freedom and "hog the entire chocolate stash if he wanted to," Mason explains, "and he does not. We were shocked." But even more surprising was that, over many tests with many rats, after releasing the trapped rat, the free rat shared the chocolate with him, letting the released rat have, on average, one and a half of the five chips. According to Bartal, this kind of generosity would be impressive even for humans.

After Bartal and Mason's rat experiments were reported in the popular press, people reached out to them to share their own stories about rats and other animals behaving compassionately. One especially poignant story—from a biology teacher—described what happened when one of his two classroom rats became paralyzed. The healthy rat didn't abandon his buddy. Every day he pulled the paralyzed rat to the water and food bowls.

Despite human efforts throughout history to eradicate rats, they are still here—so clearly, they are a successful species. Perhaps Darwin was right: if rats are as sympathetic a species as these experiments suggest, this might explain why they have flourished.

Mirror, Mirror: Who's the Most Empathetic of All?

It's not unusual to get choked up when watching a sad movie. Similarly, we often wince when we see someone injure themselves, as if we ourselves were somehow feeling their pain. Imaging shows that this phenomenon of "feeling another's pain" is located in a specific region of the brain called the anterior cingulate cortex. This part of the brain is activated both when we feel our own pain *and* when we merely observe someone else's pain.

Scientist Christian Keysers discovered that the rat brain is wired pretty much the same way as the human brain. Keysers designed an experiment in which he administered a mild electroshock to a rat while another rat watched. The shocked rat froze in place, which is a typical rat reaction to feeling fear. For example, if a rat sees a cat, he usually freezes. So, the rat freezing in response to a shock didn't surprise the scientists. But what did surprise them was the observer rat's response. He froze, too, as if he were receiving the shock.

During the experiment, the scientists used imaging technology to look at the rats' brains and saw that the same region in a rat's brain—the cingulate cortex—is activated both when he experiences pain and when he sees another rat experience pain. Furthermore, they learned that specific brain cells in the rat's cingulate cortex—called mirror neurons—fire under both conditions. For some time now, scientists have suspected that mirror neurons are associated with empathy, and Keysers believes his experiment supports this theory.

Sadly, experiments that look for empathy in rats sometimes call into question the empathy of human researchers conducting those experiments. Hopefully, now that so many studies have demonstrated that these sensitive, intelligent rodents experience empathy, we can start treating them more humanely . . . or more like a rat would.

Mice Are Nice, Too

Like rats, mice respond to the pain of other mice. They squirm when seeing another mouse in discomfort, and they get even more agitated if the mouse is familiar to them—a friend or family member. Also similar to their rat cousins, mice will rescue a trapped cage mate. Japanese scientists conducted an experiment in which one mouse was trapped in a transparent tube with a small breathing hole at one end and a paper lid on the other. The mouse was positioned so that her head was at the breathing-hole end.

Like rats, the free mice released their trapped cage mates: they gnawed at the paper lid until the trapped mouse was free. Scientists investigated whether mice were only interested in freeing familiar mice, but the free mice released the trapped mice regardless of familiarity. To make sure the mice weren't simply enjoying chewing on the paper lids, the researchers presented them with paper-lid-sealed empty tubes, but the mice ignored them.

Bioethicist Jessica Pierce relates another story that highlights the mouse's capacity for compassion. CeAnn Lambert, the founder of the Indiana Coyote Rescue Center, told Pierce about two baby mice she once discovered trapped in a sink. Lambert gave the mice water in a small lid. One mouse immediately scurried over and took a drink, but the other seemed too weak to get himself to the water. The strong mouse immediately found a tiny scrap of food and used it to lure the weak mouse closer and closer to the water, until he was close enough to take a drink. It seems that kindness sometimes comes in the smallest packages.

Like rats, mice seem to understand one another's moods, including when a fellow mouse is in pain.

Epilogue

The stories presented in this book—about a giant humpback whale protecting a marine biologist from a shark, a tiny mouse helping his weakened companion take a drink of water, and so many others—make a compelling case for animals' capacity for kindness. These remarkable stories, and similar accounts of animal behavior, should change the way we regard and treat the creatures with whom we share our world.

However, there are still many people, including some in the scientific community, who brush off such stories, even when reported by experienced marine biologists, primatologists, and other experts in the field. The established perspective holds that if a behavior was only witnessed once, just a few times, or did not occur under controlled experimental conditions, it is merely anecdotal.

But as political scientist Raymond Wolfinger astutely observed: "the plural of anecdote is data."

During the past two decades, stories about empathetic and altruistic animal behavior have become increasingly common. Based on the pace of new discoveries in this arena, my guess is that these stories will soon become data, either as a result of their sheer abundance or because they inspire further scientific studies. Many reports about animal behavior have already galvanized scientists to undertake more rigorous research that often ends up supporting the conclusions suggested by the anecdotes.

While skeptics might need to wait for further proof that animals are capable of empathy and altruism, the rest of us do not. Instead, we could simply decide that stories like those presented here—and others that are regularly reported in both the popular press and academic journals—are enough to justify changing our perspectives and behaviors. Recognizing that animals are capable of compassion should inspire the same in us.

So let's do what we can to make the world a kinder place for animals—and ourselves—by better addressing habitat destruction, factory farming, pet abandonment, and other forms of cruelty to animals. In this way, we would be following the wise advice of poet and philosopher Ralph Waldo Emerson, who wrote, "You cannot do a kindness too soon, for you never know how soon it will be too late."

—Belinda Recio

About the Author

Belinda Recio is a recipient of the Humane Society's Award for Innovation in the Study of Animals and Society. She has developed award-winning science curricula for educational television, museums, and publishers and has authored books on subjects ranging from spiritual traditions to animal behavior. Her most recent book, *Inside Animal Hearts and Minds*, on animal emotion and intelligence, was published by Skyhorse Publishing in 2017.

Belinda is also the owner and director of True North Gallery in Hamilton, Massachusetts (www.truenorthgallery.net), where she exhibits art that connects people with animals and nature.

Notes

All links and Web addresses were checked and verified to be correct at the time of publication. Because of the dynamic nature of the Web, some addresses and links may have changed since publication and may no longer be valid.

INTRODUCTION

Annie Dillard, "A Writer in the World," in *Abundance: Narrative Essays Old and New* (New York: HarperCollins, 2016 EPub edition), 105.

Frans de Waal, "Your Dog Feels as Guilty as She Looks: Animals are no less emotional than we are." *New York Times,* Sunday Review: Opinion, www.nytimes.com, March 8, 2019, https://www.nytimes.com/2019/03/08/opinion/sunday/emotions-animals-humans.html.

CHAPTER 1: WHALE

A Beautiful Question

Sarah Gibbens, "Whale Allegedly Protects Diver from Shark, but Questions Remain," *National Geographic*, www.nationalgeographic.com, January 11, 2018, https://www.nationalgeographic.com/news/2018/01/whale-protects-diver-from-shark-video-spd/.

Nan Hauser, interviewed by Ari Shapiro, National Public Radio "All Things Considered," radio programming transcript, "How a Whale Saved a Marine Biologist from a Shark," posted on NPR "Animals," www.npr.org, January 12, 2018, https://www.npr.org/2018/01/12/577713381/how-a-whale-saved-a-marine-biologist-from-a-shark.

Brandon Specktor, "This Humpback Whale Saved a Woman's Life, But Probably Not on Purpose," Live Science, www.livescience.com, January 09, 2018, https://www.livescience.com/61380-humpback-whale-saves-diver-video.html.

A Compassionate Instinct?
Robert L. Pitman, Volker B. Deecke, Christine M. Gabriele, et al., "Humpback whales interfering when mammal-eating killer whales attack other species: Mobbing behavior and interspecific altruism?," *Marine Mammal Science* 3, no. 1 (July 20, 2016): 7–58, doi:10.1111/mms.12343, https://onlinelibrary.wiley.com/doi/full/10.1111/mms.12343.

Jason Bittel, "Why Humpback Whales Protect Other Animals from Killer Whales," *National Geographic*, www.nationalgeographic.com, August 8, 2016, https://www.nationalgeographic.com/news/2016/08/humpback-whales-save-animals-killer-whales-explained/.

Joshua Howgego, "I saw humpback whales save a seal from death by killer whale," *New Scientist*, www.newscientist.com, October 15, 2016, https://www.newscientist.com/article/mg23230950-700-i-saw-humpback-whales-save-a-seal-from-death-by-killer-whale/.

Beluga to the Rescue
Mary Ann Potts and Laura Buckley, "Beluga Whale Saves Drowning Diver," *National Geographic* "Adventure Blog," www.nationalgeographic.com, August 3, 2009, https://www.nationalgeographic.com/adventure/adventure-blog/2009/08/03/mila-to-the-rescue-beluga-whale-saves-drowning-diver/.

"Beluga whale 'saves' diver," *The Telegraph*, www.telegraph.co.uk, July 29, 2009, https://www.telegraph.co.uk/news/newstopics/howaboutthat/5931345/Beluga-whale-saves-diver.html.

"Amazing rescue: Drowning diver saved by beluga whale," *The Mirror*, www.mirror.co.uk, July 29, 2009, https://www.mirror.co.uk/news/weird-news/amazing-rescue-drowning-diver-saved-409479.

The Cells That Make Us . . . Whale?
"The 'Human Neuron,' not so special after all?," *The Cellular Scale Blog*, www.cellularscale.blogspot.com, January 26, 2012, http://cellularscale.blogspot.com/2012/01/human-neuron-not-so-special-after-all.html.

Andy Coghlan, "Whales boast the brain cells that 'make us human'," *New Scientist*, www.newscientist.com, November 27, 2006, https://www.newscientist.com/article/dn10661-whales-boast-the-brain-cells-that-make-us-human/.

Patrick R. Hof and Estel Van Der Gucht, "Structure of the cerebral cortex of the humpback whale, *Megaptera novaeangliae* (Cetacea, Mysticeti, Balaenopteridae)," *The Anatomical Record* 290, no. 1 (January 23, 2007): 1–31, doi:10.1002/ar.20407, https://anatomypubs.onlinelibrary.wiley.com/doi/full/10.1002/ar.20407.

A Whale of Gratitude

Peter Fimrite, "Daring rescue of whale off Farallones /Humpback nuzzled her saviors in thanks after they untangled her from crab lines, diver says," SFGate, www.sfgate.com, December 14, 2005, https://www.sfgate.com/bayarea/article/Daring-rescue-of-whale-off-Farallones-Humpback-2557146.php.

"Connection," YouTube video, 2:13, posted by White Shark Video, December 2, 2012. https://www.youtube.com/watch?v=4w8GjX7pyfM.

CHAPTER 2: PARROT

More than Just a Mimic

Irene M. Pepperberg, *Alex & Me: How a Scientist and a Parrot Discovered a Hidden World of Animal Intelligence—and Formed a Deep Bond in the Process* (New York: HarperCollins, 2008), 149.

Associated Press, "Parrot gets award for warning about choking tot," *NBC News*, www.ncbbews.com, March 24, 2009, http://www.nbcnews.com/id/29858577/ns/us_news-wonderful_world/t/parrot-gets-award-warning-about-choking-tot/#.XZpLmi2ZOH5.

"Parrot saved toddler's life with warning," *The Telegraph*, www.telegraph.co.uk, March 25, 2009, https://www.telegraph.co.uk/news/worldnews/northamerica/usa/5048970/Parrot-saved-todlers-life-with-warning.html.

Need a Guard Dog? Consider a Parrot Instead!

"Pet parrot Wunsy saves owner from park attack in London," *BBC News* online, www.bbc.com, April 10, 2014, https://www.bbc.com/news/uk-england-london-26967945.

"Pet parrot saves owner from sex attack," News.com.au, www.news.com.au, April 12, 2014, https://www.news.com.au/technology/science/pet-parrot-saves-owner-from-sex-attack/news-story/c89fcb9713acd6dcad5e297e77418dcd.

Emma Glanfield, "Peck on someone your own size!," *Daily Mail*, www.dailymail.co.uk, April 10, 2014, https://www.dailymail.co.uk/news/article-2601364/Peck-size-African-Grey-parrot-saves-owner-mugger-pushed-ground-squawking-flapping-fled.html.

The Human-Parrot Bond
Joanna Burger, *The Parrot Who Owns Me: The Story of a Relationship* (PLACE: Random House, 2001), Kindle edition, location 59.

CHAPTER 3: GORILLA

For the Greater Good
Ker Than, "Gorilla Youngsters Seen Dismantling Poachers' Traps—A First," *National Geographic,* www.nationalgeographic.com, July 18, 2012, https://www.nationalgeographic.com/news/2012/7/120719-young-gorillas-juvenile-traps-snares-rwanda-science-fossey/.

Bob Yirka, "Gorillas filmed performing amazing feat of intellectual ability," Phys Org, www.phys.org, July 23, 2012, https://phys.org/news/2012-07-gorillas-amazing-feat-intellectual-ability.html.

Bec Crew, "Young Gorillas Have Learnt How to Dismantle Poachers' Traps in the Wild," Science Alert, www.sciencealert.com, January 26, 2016, https://www.sciencealert.com/young-gorillas-seen-dismantling-poachers-traps-for-the-first-time.

Nurturing Nature: Binti Jua
Julia Jacobo, "Gorilla Carries 3-Year-Old Boy to Safety After He Fell Into Enclosure in 1996 Incident," *ABC News* online, www.abcnews.com, May 30, 2016, https://abcnews.go.com/US/gorilla-carries-year-boy-safety-fell-enclosure-1996/story?id=39479586.

Associated Press, "Gorilla at an Illinois Zoo Rescues a 3-Year-Old Boy," *New York Times*, www.nytimes.com, August 17, 1996, https://www.nytimes.com/1996/08/17/us/gorilla-at-an-illinois-zoo-rescues-a-3-year-old-boy.html.

"15 Years Ago Today: Gorilla Rescues Boy Who Fell in Ape Pit," *CBS News Chicago* online, www.chicago.cbslocal.com, August 16, 2011, based on reporting by Regine Schlesinger of WBBM Newsradio, https://chicago.cbslocal.com/2011/08/16/15-years-ago-today-gorilla-rescues-boy-who-fell-in-ape-pit/.

Jambo

Robert John Young, "Harambe the Gorilla Put Zoo In A Lose-Lose Situation—By Being Himself," The Conversation, www.theconversation.com, May 31, 2016, https://theconversation.com/harambe-the-gorilla -put-zoo-in-a-lose-lose-situation-by-being-himself-60278.

Stephen Messenger, "Man 'Forever Thankful' To Gorilla Who Saved His Life As A Child," The Dodo, www.thedodo.com, January 21, 2016, https://www.the dodo.com/gorilla-saves-child-video-1565669019.html.

Frown, Sad, Trouble

Francine Patterson and Wendy Gordon, "The Case for the Personhood of Gorillas," in *The Great Ape Project*, eds. Paola Cavalieri and Peter Singer (New York: St. Martin's Griffin, 1993), 65.

Patterson and Gordon, "The Case for the Personhood of Gorillas," 76.

"Koko Responds to a Sad Movie," video, 2:00, videographer Ron Cohn, posted by Gary Stanley, "Do Gorillas Feel Empathy?," *KokoFlix Blog*, August 7, 2014, Koko.org—The Gorilla Foundation, www.koko.org, https://www.koko.org/kokoflix-video-blog/3860/do-gorillas-feel-empathy/.

CHAPTER 4: DOG

The Crow Pass Guide Dog

Hilary Bird, "'He gave me the motivation to get up:' Woman recounts rescue by Nanook the husky," *CBC News* online, www.cbc.ca, July, 5, 2018, https://www.cbc.ca/news/canada/north/husky-rescue-deaf-woman -alaska-1.4734018.

Heather Hintze, "Heroic husky made honorary search & rescue dog," *KTVA News* online, www.ktva.com, November 8, 2018, https://www.ktva.com/story/39450303/heroic-husky-made-honorary-…GmGXi -mRaVrfntoKi1Q1Y-S2r159XwCzXc3-agXE#.W-UoFon2ac8.facebook.

Patti Singer, "RIT hiker shares more details of Alaska rescue by heroic husky," *Democrat and Chronicle*, www .democratandchronicle.com, June 27, 2018, https://www.democratandchronicle.com/story/news/2018/06/27 /amelia-milling-hiker-rescue-husky-nanook-rit-alaska-troopers/735632002/.

"Meet the heroic dog who saved an injured and deaf hiker," *KTVA News* online, www.KTVA.com, June 26, 2018, https://www.ktva.com/story/38517370/meet-the-heroic-dog-who-saved-an-injured-and-deaf-hiker.

Returning the Favor

Courier Mail, "Doberman saves toddler," *Adelaide Now* "The Advertiser" online, www.adelaidenow.com.au, News Corps Australia, October 30, 2007, https://www.adelaidenow.com.au/news/national/doberman-saves -toddler/news-story/3b4861fab 6c3e3a171017fc2e56780bb?sv=eb87b1c5e83cddcf d1692724a26db527.

Nicolette Weet, "A mother brings home a new dog that saves her child's life," KiwiReport, www.kiwireport .com, September 25, 2017, http://www.kiwireport.com/mother-brings-home-new-dog-saves-childs-life/.

A Guardian Angel

"Boy calls dog who fought off cougar his 'guardian'," *CBC News* online, www.cbc.ca, January 3, 2010, https: //www.cbc.ca/news/canada/british-columbia/boy-calls-dog-who-fought-off-cougar-his-guardian-1.874080.

Mike Celizic, "Hero dog saves boy, 11, from cougar attack," *Today Blog*, January 5, 2010, https://www.today .com/pets/hero-dog-saves-boy-11-cougar-attack-2D80555490.

Not Just a Warm and Fuzzy Feeling

Johns Hopkins University, Jill Rosen, Office of Communications, "What Would Your Dog Do to Help If You Were Upset? Quite a Bit, Study Finds," July 24, 2018, https://releases.jhu.edu/2018/07/24/what-would -your-dog-do-to-help-if-you-were-upset-quite-a-bit-study-finds/.

Emily M. Sanford, Emma R. Burt, Julia E. Meyers-Manor, "Timmy's in the well: Empathy and prosocial helping in dogs," *Learning & Behavior* 46, no. 4 (December 2018): 374–86. https://doi.org/10.3758 /s13420-018-0332-3.

CHAPTER 5: LION

When the Lion Lies Down with the Baby Oryx

Natalie Wolchover, "Did Lioness Really Befriend Baby Antelope?," Live Science, www.livescience.com, October 10, 2012, https://www.livescience.com/34279-lioness-baby-antelope-kob.html.

Marc Lacey, "5 Little Oryxes and the Big Bad Lioness of Kenya," *New York Times*, www.nytimes.com, October 12, 2002, https://www.nytimes.com/2002/10/12/world/5-little-oryxes-and-the-big-bad-lioness -of-kenya.html.

Saba Douglas-Hamilton, "Heart of a Lioness: Letters from experts commenting on Kamunyak's story," *Saba Douglas-Hamilton Blog*, January 19, 2013, www.sabadouglashamilton.com, https://sabadouglashamilton.com /heart-of-a-lioness/.

Rescued by Lions?

Joy Adamson, *Born Free: A Lioness of Two Worlds* (New York: Pantheon Books, 1987), 108.

"Kidnapped girl 'rescued' by lions," *BBC News* online, www.news.bbc.co.uk, June 22, 2005, http://news.bbc.co.uk/2/hi/africa/4116778.stm.

Associated Press, "Ethiopian girl reportedly guarded by lions," *NBC News* online, www.nbcnews.com, updated June 21, 2005, http://www.nbcnews.com/id/8305836/ns/world_news-africa/t/ethiopian-girl-reportedly-guarded-lions/#.XZn6Uy2ZOH5.

Craig Packard, email exchanges with the author, May 30 and 31, 2019.

The Leonine Sisterhood

Reuters, "For Female Lions, Democracy Rules," *New York Times*, www.nytimes.com, July 31, 2001, https://www.nytimes.com/2001/07/31/science/for-female-lions-democracy-rules.html.

CHAPTER 6: MONKEY

Monkey Paramedics

Associated Press, "Life-saving monkey steps in to aid a friend after electric shock," *Boston Globe*, www.bostonglobe.com, December 26, 2014, https://www.bostonglobe.com/news/world/2014/12/26/monkey-gives-first-aid-friend-after-electric-shock/tKM4g0B c7CyBNgYbJkBNJJ/story.html.

"Monkey saves dying monkey at Kanpur railway station in India," YouTube video, 3:34, posted by News Hour India, December 22, 2014, https://www.you tube.com/watch?v=ulg1Imcavew.

Brian Clark Howard, "Was Monkey Actually Trying to Revive Shocked Companion," *National Geographic*, www.nationalgeographic.com, December 23, 2014, https://www.nationalgeographic.com/news/2014/12/141223-monkey-cpr-revives-kanpur-india-video-animals/.

Monkey Midwifery

University of Chicago, "Scientists learn about humans' Machiavellian intelligence by looking at the behavior of monkeys," October 5, 2019, University of Chicago News Office, www-news.uchicago.edu, http://www-news.uchicago.edu/releases/07/071025.monkeys.shtml.

Michael Marshall, "'It's a boy!' Monkey midwife delivers baby," *New Scientist* "Zoologger," www.new scientist.com, February 8, 2013, https://www.newsci entist.com/article/dn23149-zoologger-its-a-boy -monkey-midwife-delivers-baby/.

Matt Walker, "The Monkey that Became a Midwife," *BBC News* online, www.bbc.com, October 6, 2014, http://www.bbc.com/earth/story/20141006-the-monkey-that-became-a-midwife?ocid=twert.

Melissa Hogenboom, "The monkeys that act as midwives," *BBC News* online, www.bbc.com, April 16, 2016, http://www.bbc.com/earth/story/20160414-the-monkeys-that-act-as-midwives.

Wenshi Pan, Tieliu Gu, Yue Pan, et al., "Birth intervention and non-maternal infant-handling during parturition in a nonhuman primate," *Primates* 55, no. 4 (May 24, 2014): 483–88.

It Is (Biologically) Better to Give than to Receive

Jules H. Masserman, Stanley Wechkin, and William Terris, "Altruistic Behavior in Rhesus Monkeys," *American Journal of Psychiatry* 121, (1964): 584–85.

Michael Marshall, "Monkeys chill out just from seeing their friends being groomed," *New Scientist*, www .newscientist.com, December 12, 2018, https://www.newscientist.com/article/2187943-monkeys-chill-out -just-from-seeing-their-friends-being-groomed/#ixzz61bBL0v7k.

Juliette M. Berthier and Stuart Semple, "Observing grooming promotes affiliation in Barbary macaques," *Proceedings of the Royal Society B* 285 (December 12, 2018), doi.org/10.1098/rspb.2018.1964, https://royal societypublishing.org/doi/10.1098/rspb.2018.1964.

CHAPTER 7: DOLPHIN

The Good Citizens of the Sea

"Dolphins save swimmers from shark," *CBC News* online, www.cbc.ca, November 24, 2004, https://www .cbc.ca/news/world/dolphins-save-swimmers-from-shark-1.513716.

Sam Jones, "Dolphins save swimmers from shark," *The Guardian*, www.theguardian.com, November 23, 2004, https://www.theguardian.com/science/2004/nov/24/internationalnews.

Mysterious Minds

Maddalena Bearzi and Craig Stanford, *Beautiful Minds: The Parallel Lives of Great Apes and Dolphins* (Cambridge, MA: Harvard University Press, 2010): 23–27.

They Had His Back

Emily Thomas, "This Swimmer Noticed A Shark Was Following Him; The Dolphins Noticed, Too," Huffington Post, www.huffpost.com, April 26, 2014, Updated April 29, 2014, https://www.huffpost.com /entry/swimmer-saved-by-dolphins_n_5215041.

Across the Species Divide

Karen B. London, "Dolphins Save Dog From Drowning," The Bark, www.thebark.com, March 2011, Updated July 2016, https://thebark.com/content/dolphins-save-dog-drowning.

Carly Schwartz, "Dog Rescued By Dolphins In Florida Canal," Huffington Post, www.huffpost.com, February 24, 2011, Updated December 6, 2017, https://www.huffpost.com/entry/dog-rescued-by -dolphins-video_n_828019.

Kindred Spirits in the Sea

Amanda Pachniewska, "List of Animals That Have Passed the Mirror Test," Animal Cognition, www .animalcognition.org, http://www.animalcognition.org/2015/04/15/list-of-animals-that-have-passed-the -mirror-test/.

Stephanie L. King and Vincent M. Janik, "Bottlenose dolphins can use learned vocal labels to address each other," *Proceedings of the National Academy of Sciences* 110, no. 32 (July 22, 2013): 13216–21, https://www .pnas.org/content/early/2013/07/17/1304459110.

Stephanie L. King, Laela S. Sayigh, Randall S. Wells, et al., "Vocal copying of individually distinctive signature whistles in bottlenose dolphins," *Proceedings of the Royal Society B* 280 (April 22, 2013), doi .org/10.1098/rspb.2013.0053, https://royalsocietypub lishing.org/doi/10.1098/rspb.2013.0053.

Christie Wilcox, "Do Stoned Dolphins Give 'Puff Puff Pass' A Whole New Meaning?" *Discover Magazine*, www.discovermagazine.com, December 30, 2013, https://www.discovermagazine.com/planet-earth/do -stoned-dolphins-give-puff-puff-pass-a-whole-new-meaning?utm_source=twitterfeed&utm_medium =twitter#.UsL7iGRDv2F.

Mark H. Deakos, Brian K. Branstetter, Lori Mazzuca, et al., "Two unusual interactions between a bottlenose dolphin (*Tursiops truncatus*) and a humpback whale (*Megaptera novaeangliae*) in Hawaiian waters," *Aquatic Mammals* 36, no. 2 (October 13, 2010): 121–28, doi:10.1578/AM.36.2.2010.121.

Charles Q. Choi , "10 Animals That Use Tools," Live Science, www.livescience.com, December 14, 2009, https://www.livescience.com/9761-10-animals-tools.html.

Jeffrey Moussaieff Masson and Susan McCarthy, *When Elephants Weep: The Emotional Lives of Animals* (New York: Dell Publishing, 1995), 127.

T. G. Leighton , G. H. Chua, and P. R. White, "Do dolphins benefit from nonlinear mathematics when processing their sonar returns?" *Proceedings of the Royal Society A* 468 (July 18, 2012), doi.org/10.1098 /rspa.2012.0247, https://royalsocietypublishing.org/doi/10.1098/rspa.2012.0247.

CHAPTER 8: CAT

The Order of the Blue Tiger

"My Cat Saved My Son," YouTube video, 00:56, posted by Roger Triantafilo, May 14, 2014, https://www.youtube.com/watch?v=C-Opm9b2WDk.

Victoria Heuer, "2015's Dog of the Year Award Goes to Tara the Hero Cat," PetMD, www.petmd.com, June 22, 2015, https://www.petmd.com/news/lifestyle-entertainment/2015s-dog-year-award-goes-tara-hero -cat-32838.

Jessica Hullinger, "Hero cat proves felines can be as protective as dogs," *New York Post*, www.nypost.com, June 1, 2014, https://nypost.com/2014/06/01/hero-cat-proves-felines-may-be-more-protective-than-dogs/.

Tribune Wire Reports, "Dog attacks boy, cat attacks dog; award goes to cat," *Chicago Tribune* www.chicago tribune.com, June 19, 2015, https://www.chicagotribune.com/lifestyles/pets/ct-hero-cat-award-20150619 -story.html.

Ruth Brown, "Bakersfield heroes honored Friday, including hero cat," *Bakersfield Californian*, www.bakers field.com, September 26, 2014, https://www.bakersfield.com/news/bakersfield-heroes-honored-friday -including-hero-cat/article_a1e3904f-6924-5407-95bf-ea86f59c0f0e.html.

The Alarm Cat

"Cat saves family from carbon monoxide poisoning," YouTube video, 2:16, posted by FOX 8 News Cleveland, May 4, 2016, https://www.youtube.com/watch?v=ja1VVoTRBao.

Karen Tietjen, "Hero Cat Saves Humans From Carbon Monoxide Poisoning," I Heart Cats, www .iheartcats.com, https://iheartcats.com/hero-cat-saves-humans-from-carbon-monoxide-poisoning/.

Cats in the Coal Mine

Natalie Cornish, "One in ten dog owners say their pet has saved their life," *Country Living*, www.CountryLiving.com, November 22, 2018, https://www.countryliving.com/uk/wildlife/pets/a25258550/carbon-monoxide-poisoning-pets/.

The Feline First Responder

Kristen A. Kruger, telephone discussions with the author, June 3 and 4, 2019.

Kathryn Ross, "Wellsville cat featured in USA Today for saving owner," *Wellsville Daily Reporter*, www.wellsvilledaily.com, December 23, 2010, https://www.wellsvilledaily.com/x934178835/Wellsville-cat-featured-in-USA-Today-for-saving-owner.

I. Merola, M. Lazzaroni, S. Marshall-Pescini, et al., "Social referencing and cat–human communication," *Animal Cognition* 18, no. 3 (May 2015): 639–48.

CHAPTER 9: PIG

When Pigs Fly: How a Potbellied Pig Surprised the World

Michael A. Fuoco, "LuLu the heroic pig now known worldwide," *Post-Gazette*, www.old-post-gazette.com, April 9, 2002, http://old.post-gazette.com/neigh_west/20020409lulu0409p1.asp.

"Kelly's Curiosities 2: Lulu the Hero Pig," YouTube video, 2:32, posted by Nooj, July 17, 2015, https://www.youtube.com/watch?time_continue=4&v=TNhBVesKkt0.

Michael A. Fuoco, "Oinking for help: Pot-bellied pig saves owner's life by lying in front of a car," *Post-Gazette*, www.old-post-gazette.com, October 10, 1998, http://old.post-gazette.com/regionstate/19981010pig2.asp.

Pulling Her Weight

"Pig saves her owner's bacon," *BBC News* online, www.news.bbc.co.ok, March 9, 2000, http://news.bbc.co.uk/2/hi/uk_news/wales/670625.stm.

"If You Were a Pig, You Would Have This Figured Out by Now"

"The Hidden Lives of Pigs," PETA, www.peta.org, https://www.peta.org/issues/animals-used-for-food/factory-farming/pigs/hidden-lives-pigs/.

"About Pigs," Humane Society of the United States, www.humanesociety.org, May 2015, https://www
.humanesociety.org/sites/default/files/docs/about-pigs.pdf.

Christine Dell'Amore, "Pigs recorded using tools for the first time," *National Geographic*, www.national
geographic.com, October 4, 2019, https://www.nationalgeographic.com/animals/2019/10/first-tool-use-pigs
-visayan-endangered/.

Ingrid Newkirk, "9 Ways Pigs Are Smarter than Your Honor Student," Huffington Post, www.huffpost
.com, April 15, 2014, https://www.huffpost.com/entry/9-ways-pigs-are-smarter-t_b_5154321.

Felicity Muth, "Can Pigs Empathize?," *Scientific American Blog*, www.scientificamerican.com, January
13, 2015, https://blogs.scientificamerican.com/not-bad-science/can-pigs-empathize/?redirect=1.

Lori Marino, "Thinking Pigs: A Comparative Review of Cognition, Emotion, and Personality in *Sus
domesticus*," Human Society Institute for Science and Policy: Animal Studies Repository, www.animalstudies
repository.org, 2015, https://animalstudiesrepository.org/cgi/viewcontent.cgi?article=1042&context
=acwp_asie.

Sy Montgomery, *The Good Good Pig: The Extraordinary Life of Christopher Hogwood* (New York: Ballantine
Books, April 17, 2007), Kindle edition, location 1421.

CHAPTER 10: BEAR
The Benevolent Bear
Else Poulsen and Stephen Herrero, *Smiling Bears: A Zookeeper Explores the Behaviour and Emotional Life of
Bears* (Vancouver, BC: Greystone Books, 2009), Kindle edition, locations 409, 435, 439, 455–72.

Kinder than the Average Bear
"Crow Rescue," YouTube video, 2:12, posted by Aleksander Medveš, June 21, 2014, https://www.youtube
.com/watch?v=gJ_3BN0m7S8.

Stephen Luntz, "Bear Saves Crow from Drowning," IFLScience!, www.iflscience.com, August 3, 2014,
https://www.iflscience.com/plants-and-animals/bear-saves-crow-drowning/.

Guardians of the Forest?
Amir Vera and Samira Said, "A boy who was lost in the woods says a bear kept him company. No one can
prove it didn't happen," CNN online, www.cnn.com, last modified 3:06 p.m., January 29, 2019, video, 2:00,
produced by WTVD, https://www.cnn.com/2019/01/28/us/casey-hathaway-bear-claims/index.html.

Katherine Hignett, "NC Toddler Missing in Woods Says He Survived Thanks to a Friend—'He Hung Out with a Bear for Two Days,'" *Newsweek*, www.newsweek.com, January, 29, 2019, video, 00:37, produced by CBS 17, https://www.newsweek.com/casey-hathaway-north-carolina-missing-toddler -child-bear-1308858.

Associated Press, "Bear Falsely Accused of Stealing Girl," *Waterloo Daily Courier*, July 5, 1955, digitally clipped by BruceMount on January 27, 2019, posted on www.newspapers.com, "Ida Mae Curtis," https: //www.newspapers.com/clip/27717855/ida_mae_curtis/.

Beatrice O'Hearn, "How Scottville Came to Be and Table of Early Scottville Pioneers," February 26, 1942, Madison County History Companion "Scottville History & Links," www.ludingtonmichigan.net, http: //ludingtonmichigan.net/index.php?page=Towns/scottville.

Benjamin Kilman, *In the Company of Bears* (VT: Chelsea Green, 2013), 104.

Bear in Mind: They're a Lot Like Us
Kilman, *In the Company of Bears*, 97, 101–02.

Jennifer Vonk and Michael J. Beran, "Bears 'Count' Too: Quantity Estimation and Comparison in Black Bears (*Ursus Americanus*)," *Animal Behaviour* 84, no. 1 (July 2012): 231–38, https://www.ncbi.nlm.nih .gov/pmc/articles/PMC3398692/.

CHAPTER 11: HORSE

A Horse as Good as Kerry Gold

Fiona Boyd, "My horse saved me from a raging cow," *The Guardian*, www.theguardian.com, November 21, 2014, https://www.theguardian.com/lifeandstyle/2014/nov/21/experience-my-horse-saved-me-from-raging-cow.

Trevor Cooper, "Horse saves owner," *Horse & Hound*, www.horseandhound.co.uk, August 14 2007, https: //www.horseandhound.co.uk/news/horse-saves-owners-life-after-attack-by-cow-136679.

Rescue Reciprocity

Heather Regan White, "Stormy saves the day," *The American Press*, reported in *Houma Today* online, www .houmatoday.com, November 29, 2010, https://www.houmatoday.com/entertainment/20101129/stormy-saves -the-day.

"Stormy saved my children from wild hog," YouTube video, 2:00, posted by Cathy Leonard, February 12, 2012, https://www.youtube.com/watch?v=dwX0dYEUyJo.

Straight from the Horse's Heart

Amy Victoria Smith, Leanne Proops, Kate Grounds, et al., "Functionally relevant responses to human facial expressions of emotion in the domestic horse (*Equus caballus*)," *Biology Letters* (February 1, 2016), doi:10.1098/rsbl.2015.0907, www.royalsocietypublishing.com, https://royalsocietypublishing.org/doi/10.1098/rsbl.2015.0907.

Leanne Proops, Kate Grounds, Amy Victoria Smith, Karen McComb, "Animals Remember Previous Facial Expressions That Specific Humans Have Exhibited," *Current Biology 28* (May 7, 2018), 1428–1432.e1–e4, https://doi.org/10.1016/j.cub.2018.03.035.

CHAPTER 12: SEAL AND SEA LION

A Protective Circle of Seals

Dylan Forest, "Seals Save Life, Need Help," *Animal People Forum*, www.animalpeopeforum.org, April 1, 1999, https://newspaper.animalpeopleforum.org/1999/04/01/seals-save-life-need-help/.

The Free Library, "Each time I sank, seals nudged me back up; Swim Woman Tells How Animals Saved Her," Retrieved October 7, 2019, from https://www.thefreelibrary.com/Each+time+I+sank%2c+seals+nudged+me+back+up%3b+Swim+Woman+Tells+How...-a060393683.

An Unlikely Lifeguard

Paul Stokes, "Seal swims to rescue of drowning dog," *The Telegraph*, www.telegraph.co.uk, June 20, 2002, https://www.telegraph.co.uk/news/uknews/1397813/Seal-swims-to-rescue-of-drowning-dog.html.

"Seal saves drowning dog," *BBC News* online, www.bbc.co.uk, June 19, 2002, http://news.bbc.co.uk/2/hi/uk/england/2053299.stm.

A Sea Lion's Gift of a Second Chance

"Bridge jumper says sea lion saved him," *Phys Org*, www.phys.org, March 4, 2015, https://phys.org/news/2015-03-bridge-jumper-sea-lion.html.

A Case for Sea Lion Empathy?

Marc Bekoff, "Animal Emotions: Exploring Passionate Natures: Current interdisciplinary research provides compelling evidence that many animals experience such emotions as joy, fear, love, despair, and grief—we are not alone," *BioScience* 50, no. 10 (October 2000): 861–70, https://doi.org/10.1641/0006–3568(2000)050 [0861:AEEPN]2.0.CO;2.

Elisabetta Palagi, Federico Guillén-Salazar, and Clara Llamazares-Martínet, "Spontaneous Yawning and its Potential Functions in South American Sea Lions (*Otaria flavescens*)," *Scientific Reports* 9, no. 1 (November 2019): doi:10.1038/s41598-019-53613-4.

"Sea lion pup jumps on boat, cuddles with driver," YouTube video, 1:01, posted by CBS News, June 12, 2013, https://www.youtube.com/watch?v=QMhaKIEBa7M.

Peter Cook, Andrew Rouse, Margaret Wilson, et al., "A California sea lion (*Zalophus californianus*) can keep the beat: Motor entrainment to rhythmic auditory stimuli in a non vocal mimic," *Journal of Comparative Psychology* 127, no. 4 (November 2013): 412–27, doi:10.1037/a0032345.

"Beat Keeping in a California Sea Lion (Ronan)," YouTube video, 2:13, posted by PinnipedLab, March 31, 2013, https://www.youtube.com/watch?v=6yS6qU_w3JQ.

Jill Suttie, "Where Music and Empathy Converge in the Brain," *Greater Good Magazine*, Greater Good Science Center, www.greatergood.berkeley.edu, October 22, 2018, https://greatergood.berkeley.edu/article/item/where_music_and_empathy_converge_in_the_brain.

CHAPTER 13: ORANGUTAN

The Heart and Mind of an Orangutan

H. Lyn White Miles, "Language and the Orangutan: The Old 'Person' of the Forest" in *The Great Ape Project*, eds. Paola Cavalieri and Peter Singer (New York: St. Martin's Griffin, 1993), 42–57.

William A. Hillix and Duane M. Rumbaugh, *Animal Bodies, Human Minds: Ape, Dolphin, and Parrot Language Skills* (New York: Kluwer Academic/Plenum Publishers, 2004), 194–99.

H. Lyn White Miles, "Can Chantek Talk in Codes?" in *Anthropology: The Human Challenge*, 15th ed., eds. William A. Haviland, Harald E. L. Prins, Dana Walrath, et al. (Boston: Cengage Learning, 2014, 2017), 112–13.

H. Lyn White Miles, "Me Chantek: The development of self awareness in a signing orangutan" in *Self-Awareness in Animals and Humans: Developmental Perspectives,* eds. Sue Taylor Parker, Robert W. Mitchell, and Maria L. Boccia (New York: Cambridge University Press, 1994), 254–69.

Lyn Miles, "Chantek, the first orangutan person," filmed November 17, 2016 at TEDxUTChattanooga, YouTube video, 16:27, posted by TEDx Talks, https://www.youtube.com/watch?v=q2pisrdO2TQ.

Hannah Jaicks, "Of Primates and Personhood," *Hannah Jaicks Blog,* November 3, 2014, www.hannahjaicks .com, http://www.hannahjaicks.com/blog/2014/11/3/tsn5kr1mk97wxr3r8ghubmpqc4do19.

The Friendly Orange Apes

Cathy Hayes, "An orangutan saves a duckling from drowning at the Dublin Zoo," Irish Central, January 8, 2016, www.irishcentral.com, https://www.irish central.com/news/amazing-animals-an-orangutan -at-dublin-zoo-saves-a-duckling-from-drowning-video-123911929-237393671.

Stephen Messenger, "Captive Orangutan Sneaks Food to His Friends in Never-Before-Seen Act of Kindness," The Dodo, www.thedodo.com, April 10, 2015, https://www.thedodo.com/orangutan-gives -food-to-chimps-1084748846.html.

Sarah Macdonald, "What humans can learn from orangutans," Australian Broadcasting Corporation, October 29, 2016, www.abc.net.au, https://www.abc.net.au/news/2016-10-30/what-humans-can-learn-from -orangutans/7970808.

Rebecca Le May, "Perth Zoo orangutan has a Nicole Kidman fetish," News.com.au, www.news.com.au, November 21, 2014, https://www.news.com.au/travel/australian-holidays/perth-zoo-orangutan-has-kidman -fetish/news-story/1a52ba90f2399369b588349d84d419bd.

The Person of the Forest

"Orangutan Genome Sequenced," National Institutes of Health, "NIH Research Matters," February 7, 2011, https://www.nih.gov/news-events/nih-research-matters/orangutan-genome-sequenced.

Sarah Macdonald, "What humans can learn from orangutans," Australian Broadcasting Corporation, www.abc.net.au, October 29, 2016, https://www.abc.net.au/news/2016-10-30/ what-humans-can-learn-from-orangutans/7970808.

Leif Cocks, *Orangutans My Cousins, My Friends: A journey to understand and save the person of the forest* (CA: The Orangutan Project, 2016), Kindle edition, location 1556.

Eugene Linden, "Can Animals Think?," *Time*, www.time.com, August 29, 1999, http://content.time.com /time/magazine/article/0,9171,30198–3,00.html. "Orangutan Facts," Orangutan Conservancy, www. orangutan.com, https://www.orangutan.com/orangutans /orangutan-facts/.

CHAPTER 14: ELEPHANT

An Elephant Never Forgets

AJ Willingham, "Baby elephant is so concerned about her 'drowning' friend," CNN online "Health," www .cnn.com, Updated October 17, 2016, https://www.cnn.com/2016/10/17/health/baby-elephant-saves -man-trnd/index.html.

Stephen Messenger, "Little Elephant Rushes into River to Save Her Favorite Person from 'Drowning,'" The Dodo, www.thedodo.com, October, 13, 2016, https://www.thedodo.com/elephant-rescues-favorite -person-2044205047.html.

The Kindness of Elephants

Ralph Helfer, *The Beauty of the Beasts: Tales of Hollywood's Wild Animal Stars* (New York: Open Road Media, 2014), Kindle edition, location 1445–61.

The Amazing Elephant

Benjamin L. Hart, Lynette A. Hart, Michael McCoy, et al., "Cognitive behaviour in Asian elephants: use and modification of branches for fly switching," *Animal Behaviour* 62, no. 5 (November 2001): 839–47.

Joshua M. Plotnik, Richard Lair, Wirot Suphachoksahakun, et al., "Elephants know when they need a helping trunk in a cooperative task," *Proceedings of the National Academy of Sciences* 108, no. 12 (March 2011): 5116–21, doi:10.1073/pnas.1101765108, https://www.pnas.org/content/108/12/5116.

Amanda Pachniewska, "List of Animals That Have Passed the Mirror Test," *Animal Cognition*, www.animal cognition.org, http://www.animalcognition.org/2015/04/15/list-of-animals-that-have-passed-the-mirror-test/.

Shannon Fischer, "Elephant 'Speaks' Like a Human—Uses Trunk to Shape Sound," *National Geographic*, www.nationalgeographic.com, November 23, 2012, https://www.nationalgeographic.com/news/2012 /11/121102-korean-speaking-elephant-talk-human-science-weird-animals/#close.

Christian T. Herbst, Angela S. Stoeger, Roland Frey, et al., "How Low Can You Go? Physical Production Mechanism of Elephant Infrasonic Vocalizations," *Science* 337, no. 6094 (August 2012): 595–99.

Barbara J. King, *How Animals Grieve* (Chicago: University of Chicago Press, 2013), 52–63.

CHAPTER 15: HIPPOPOTAMUS

The River Crossing Guards

"Mara River Crossing: Life and death are on parade at the most reliably perilous site of 'the Great Migration,'" contributed by brf2001, Atlas Obscura, www.atlasobscura, https://www.atlasobscura.com/places/mara-river-crossing.

Gill Gifford, "Hero hippo to the rescue," *The Star* Late Edition, December 2, 2010, posted on Pressreader, https://www.pressreader.com/south-africa/the-star-late-edition/20101202/282995396276566.

"The helpful hippo that rescues helpless animals," *Metro News*, www.metro.co.uk, November 10, 2010, https://metro.co.uk/2010/11/10/the-helpful-hippo-who-rescues-other-animals-from-mara-river-577284/.

Leon Watson, "Hip hip hooray! Hippos fight off crocodile attacking a gnu and escort the animal to safety," *Daily Mail*, www.dailymail.co.uk, March 11, 2014, https://www.dailymail.co.uk/news/article-2578257/Hip-hip-hooray-Hippos-fight-crocodile-attacking-gnu-escort-animal-safety.html.

"Heroic hippo saves wildebeest from jaws of crocodile," *Metro News*, www.metro.co.uk, March 12, 2014, https://metro.co.uk/2014/03/12/vadim-onishchenko-hero-hippo-saves-wildebeest-from-jaws-of-crocodile-4539317/.

Hippopotamus Empathy

University of California–Berkeley, Media Relations, Robert Sanders, "UC Berkeley, French scientists find missing link between the whale and its closest relative, the hippo," January 24, 2005, UCBerkeleyNews, www. berkeley.edu, https://www.berkeley.edu/news/media/releases/2005/01/24_hippos.shtml.

University of Calgary, "Is the Hippopotamus the Closest Living Relative to the Whale?," ScienceDaily, www.sciencedaily.com, March 19, 2009, https://www.sciencedaily.com/releases/2009/03/090318153803.htm.

L. Inman and Keith E. A. Leggett, "Observations on the response of a pod of hippos to a dead juvenile hippo (*Hippopotamus amphibius*, Linnaeus 1758)," originally from *African Journal of Ecology* (May 29, 2019):

1–3, uploaded by Victoria Inman to ResearchGate, www.researchgate.net, https://www.researchgate.net/publication/333458258_Observations_on_the_response_of_a_pod_of_hippos_to_a_dead_juvenile_hippo_Hippopotamus_amphibius_Linnaeus_1758.

"The 'Human Neuron,' not so special after all?," *The Cellular Scale Blog*, www.cellularscale.blogspot.com, January 26, 2012, http://cellularscale.blogspot.com/2012/01/human-neuron-not-so-special-after-all.html.

CHAPTER 16: CHIMPANZEE & BONOBO

Please Person Hug: **Washoe's Generosity of Heart**

Lawrence E. Johnson, *A Morally Deep World: An Essay on Moral Significance and Environmental Ethics*, 2nd ed. (Cambridge: Cambridge University Press, 1993), 27.

Roger Fouts, *Next of Kin: My Conversations with Chimpanzees* (New York: HarperCollins, 1997), 291, 136, 180.

Roger Fouts, "My Best Friend Is a Chimp," *Psychology Today*, www.psychologytoday.com, July 1, 2000, https://www.psychologytoday.com/us/articles/200007/my-best-friend-is-chimp.

What Is It Like to Be a Bonobo Thinking About a Starling?

Thomas Nagel, "What Is It Like to Be a Bat?," *The Philosophical Review* 83, no. 4 (October 1974): 435–50.

Frans B. M. de Waal, *Our Inner Ape: A Leading Primatologist Explains Why We Are Who We Are* (New York: Riverhead Books, 2005), Kindle edition, 2.

The Sisterhood of Motherhood: Bonobo Midwifery

Bob Yirka, "Bonobo females found to protect and support a female giving birth," Phys Org, www.phys.org, May 22, 2018, https://phys.org/news/2018-05-bonobo-females-female-birth.html.

Elisa Demuru, Pier Francesco Ferrari, and Elisabetta Palagi, "Is birth attendance a uniquely human feature? New evidence suggests that Bonobo females protect and support the parturient," *Evolution and Human Behavior* 39, no.5 (September 2018): 502–10.

Kindness of the Apes

Jingzhi Tan and Brian Hare, "Bonobos Share with Strangers," *PLoS ONE* 8, no. 1 (January 2, 2013): e51922, doi.org/10.1371/journal.pone.0051922, https://journals.plos.org/plosone/article?id=10.1371/journal.pone.0051922.

Felix Warneken, Brian Hare, Alicia P. Melis, et al., "Spontaneous Altruism by Chimpanzees and Young Children," *PLoS Biol* 5, no. 7 (June 26, 2007): e184, doi:10.1371/journal.pbio.0050184, https://journals.plos .org/plosbiology/article?id=10.1371/journal.pbio.0050184.

Nicolas Claidière, Andrew Whiten, Mary C. Mareno, et al., "Selective and contagious prosocial resource donation in capuchin monkeys, chimpanzees and humans," *Scientific Reports* 5, Article no. 07631 (January 6, 2015), doi.org/10.1038/srep07631, https://www.nature.com/articles/srep07631.

Christine E. Webb, Teresa Romero, Becca Franks, et al., "Long-term consistency in chimpanzee consolation behaviour reflects empathetic personalities," *Nature Communications* 292, no. 8 (August 2017): 1–8, doi:10.1038/s41467-017-00360-7, https://www.nature.com/articles/s41467-017-00360-7.

CHAPTER 17: CROW

Under His Wing

Linda Cole, "The True Story of a Wild Crow That Saved a Kitten," Canidae Pet Food Company Blog, April 26, 2102, www.canidae.com, https://www.canidae.com/blog/2012/04/true-story-of-wild-crow-that-saved/.

"Crow and Kitten Are Friends," YouTube video, 7:29, PAX TV Network "Miracle Pets," posted by Ozricus, June 2, 2007, https://www.youtube.com/watch?v=1JiJzqXxgxo.

Laura Moss, "Opposites attract: The kitten raised by a crow," Mother Nature Network, www.mnn.con, October 24, 2011, https://www.mnn.com/family/pets/stories/opposites-attract-the-kitten-raised-by-a-crow.

The Feathered Apes

Helen M. Ditz and Andreas Nieder, "Neurons selective to the number of visual items in the corvid songbird endbrain," *Proceedings of the National Academy of Sciences* 112, no. 25 (June 23, 2015): 7827–32, doi:10.1073/ pnas.1504245112, http://www.pnas.org/content/112/25/7827.

Gavin R. Hunt and Russell D. Gray, "The crafting of hook tools by wild New Caledonian crows," *Proceedings of the Royal Society B* 271, no. 3 (February 7, 2004), doi:10.1098/rsbl.2003.0085, https://www .ncbi.nlm.nih.gov/pmc/articles/PMC1809970/pdf/15101428.pdf.

John Marzluff and Tony Angell, *Gifts of the Crow: How Perception, Emotion, and Thought Allow Smart Birds to Behave Like Humans* (New York: Free Press, 2012): 103.

Claudia A. F. Wascher and Thomas Bugnyar, "Behavioral Responses to Inequity in Reward Distribution and Working Effort in Crows and Ravens," *PLoS One* 8, no. 2 (February 20, 2013): e56885, doi:10.1371/journal .pone.0056885, https://journals.plos.org/plosone/article?id=10.1371/journal.pone.0056885.

Jennifer Ackerman, *The Genius of Birds*, (New York: Penguin Press, 2016): 134.

CHAPTER 18: RAT AND MOUSE

Rats to the Rescue? Rethinking the Much-Maligned Rodent

Inbal Ben-Ami Bartal, Jean Decety, and Peggy Mason, "Helping a cagemate in need: empathy and pro-social behavior in rats," *Science* 334, no. 6061 (December 9, 2011): 1427–30, https://www.ncbi.nlm.nih.gov /pmc/articles/PMC3760221/.

Alka Chandna, "Rats Have Empathy, But What About the Scientists Who Experiment on Them?" The Hastings Center "Bioethics Forum," www.the hastngscenter.org, posted by Susan Gilbert, June 24, 2015, https://www.thehastingscenter.org/rats-have-empathy-but-what-about-the-scientists-who-experiment-on-them/.

Mirror, Mirror: Who's the Most Empathetic of All?

Christian Keysers, "What Makes Us Empathic?," *Psychology Today*, www.psychologytoday.com, April 11, 2019, https://www.psychologytoday.com/us/blog/the-empathic-brain/201904/what-makes-us-empathic?amp.

Maria Carrillo, Yinging Han, Filippo Migliorati, et al., "Emotional Mirrors in the Rat's Anterior Cingulate Cortex," *Current Biology* 29, no. 8 (April 22, 2019): 1301–12, www.cell.com, doi: 10.1016/j.cub.2019.03.024, https://www.cell.com/current-biology/fulltext/S0960-9822(19)30322–7.

Mice Are Nice, Too

Hiroshi Ueno, Shunsuke Suemitsu, Shinji Murakami, et al., "Helping-Like Behaviour in Mice Towards Conspecifics Constrained Inside Tubes," *Scientific Reports* 9, no. 1 (December 2019): 5817, www.nature.com, doi: 10.1038/s41598-019-42290-y, https://www.nature.com/articles/s41598-019-42290-y.

Jessica Pierce, "Mice in the Sink: On the Expression of Empathy in Animals," originally from *Environmental Philosophy* 5, no. 1 (Spring 2008): 75–96, pdf posted at www.semanticscholar.org, https://pdfs.semantic scholar.org/ed58/9c1a7dd19a7396163e5097770df 6205fa23f.pdf.

Suggested Reading

Balcombe, Jonathan P. *Second Nature: The Inner Lives of Animals*. Basingstoke: Palgrave Macmillan, 2011.

———. *Pleasurable Kingdom: Animals and the Nature of Feeling Good*. London: Macmillan, 2006.

Bearzi, Maddalena, *Beautiful Minds: The Parallel Lives of Great Apes and Dolphins*. Cambridge: Harvard University Press, 2008.

Bekoff, Marc. *Animal Passions and Beastly Virtues: Reflections on Redecorating Nature*. Philadelphia: Temple University Press, 2006.

———. *The Emotional Lives of Animals: A Leading Scientist Explores Animal Joy, Sorrow, and Empathy—and Why They Matter*. Novato, CA: New World Library, 2007.

———. *Minding Animals: Awareness, Emotions, and Heart*. New York: Oxford University Press, 2002.

Bekoff, Marc, and Jessica Pierce. *Wild Justice: The Moral Lives of Animals*. Chicago: University of Chicago Press, 2009.

de Waal, F. B. M. *The Age of Empathy: Nature's Lessons for a Kinder Society*. New York: Harmony Books, 2009.

———. *The Bonobo and the Atheist: In Search of Humanism among the Primates*. New York: W. W. Norton and Company, 2013.

———. *Mama's Last Hug: Animal Emotions and What They Tell Us about Ourselves*. New York: W.W. Norton & Company, 2019.

———. *Our Inner Ape*. New York: Riverhead Books, 2005.

Fouts, Roger. *Next of Kin: My Conversations with Chimpanzees*. New York: HarperCollins, 1997.

Heinrich, Bernd. *Mind of the Raven*: *Investigations and Adventures with Wolf-birds*. New York: Harper Collins, 2006.

Herzfeld, Chris. *Wattana: An Orangutan in Paris*. Translated by Oliver Y. Martin and Robert D. Martin. Chicago: University of Chicago Press, 2016.

King, Barbara J. *How Animals Grieve*. Chicago: University of Chicago Press, 2013.

———. *Gifts of the Crow: How Perception, Emotion, and Thought Allow Smart Birds to Behave Like Humans*. New York: Free Press, 2012.

Marzluff, John, and Tony Angell. *In the Company of Crows and Ravens*. New Haven: Yale University Press, 2005.

Masson, Jeffrey Moussaieff. *Beasts: What Animals Can Teach Us about the Origins of Good and Evil*. New York: Bloomsbury, 2014.

———. *The Pig Who Sang to the Moon: The Emotional World of Farm Animals*. New York: Ballantine Books, 2003.

Masson, Jeffrey Moussaieff, and Susan McCarthy. *When Elephants Weep: The Emotional Lives of Animals*. New York: Dell Publishing, 1995.

Montgomery, Sy. *The Good Good Pig: The Extraordinary Life of Christopher Hogwood*. New York: Ballantine Books, 2007.

———. *How to Be a Good Creature: A Memoir in Thirteen Animals*. Boston: Houghton Mifflin Harcourt, 2018.

———. *The Soul of an Octopus: A Surprising Exploration into the Wonder of Consciousness*. New York: Atria Books, 2015.

Pepperberg, Irene M. *Alex & Me: How a Scientist and a Parrot Discovered a Hidden World of Animal Intelligence—and Formed a Deep Bond in the Process*. New York: HarperCollins, 2008.

Peterson, Dale. *The Moral Lives of Animals*. New York: Bloomsbury Press, 2011.

Recio, Belinda. *Inside Animal Hearts and Minds: Bears That Count, Goats That Surf, and Other True Stories of Animal Intelligence and Emotion*. Skyhorse Publishing, 2017.

Rothenberg, David. *Thousand Mile Song: Whale Music in a Sea of Sound*. New York: Basic Books, 2008.

————. *Why Birds Sing: A Journey through the Mystery of Bird Song*. New York: Basic Books, 2005.

Rowlands, Mark. *Can Animals Be Moral?* New York: Oxford University Press, 2012.

Safina, Carl. *Beyond Words: What Animals Think and Feel*. New York: Henry Holt and Company, 2015.